Welding Practice Manual

Welder certification is not just a test; it is a true reflection of skill, knowledge, and technical preparation. This manual compiles the essential exercises for qualification in shielded metal arc welding (SMAW), TIG, MIG/MAG, flux-cored arc welding (FCAW), and oxy-fuel welding, covering materials such as carbon steel, stainless steel, cast iron, and aluminium alloys, in all joint configurations: fillet welds, butt welds, pipe-to-plate, and pipe-to-pipe connections.

Welding Practice Manual contains exercises backed by over 18,000 hours of training experience and the practical knowledge gained from preparing hundreds of welders for certification under ISO 9606 and ISO 15614 standards. However, this manual goes beyond listing exercises: it provides an innovative and structured learning approach, integrating key welding concepts that every professional should master.

This manual is designed not only to help you pass a qualification test but to understand, refine, and master welding at a professional level.

Carlos Alonso Marcos is a college instructor in Madrid and a welding coordinator in industry. He began as a welder's apprentice and has worked in the piping and high-pressure tanks sector, in land and naval transport, and the aeronautical/space industry.

I0028803

Welding Practice Manual

Carlos Alonso Marcos

Routledge
Taylor & Francis Group

LONDON AND NEW YORK

Designed cover image: Shutterstock

First published 2026
by Routledge
4 Park Square, Milton Park, Abingdon, Oxon OX14 4RN

and by Routledge
605 Third Avenue, New York, NY 10158

Routledge is an imprint of the Taylor & Francis Group, an informa business

British Library Cataloguing-in-Publication Data
A catalogue record for this book is available from the British Library

ISBN: 978-1-032-72768-4 (hbk)
ISBN: 978-1-032-69226-5 (pbk)
ISBN: 978-1-003-42248-8 (ebk)

DOI: 10.1201/9781003422488

Typeset in Times
by Apex CoVantage, LLC

Contents

Preface

In 2010, three events changed my perception of life forever: my son Darío was born, I battled melanoma, and I published this book. The story behind this manual reaffirms my belief that nothing happens by chance. I have seen enough signs and glimpses of fate to be certain of that.

From my early days as an instructor, students often asked me if I knew of any book dedicated to the practical application of welding—one they could use as a learning aid. I had spent a great deal of time and effort searching for a manual that would describe the techniques and knowledge required to achieve the skill level necessary for welding different types of joints with each of the manual welding processes, ensuring the required quality standards. However, I never found such a book.

At the start of every course, I have a habit of opening the first lesson with a theory-practice demonstration, allowing students to observe the execution, listen to my explanation, and, when possible, practise alongside me. This dynamic approach caters to all learning styles—some learn best by listening, others by watching, and others by doing. However, as useful as this method is, it soon becomes very difficult to maintain. Students in welding certification programmes come from a wide range of backgrounds, ages, experience levels, and skillsets. As a result, their learning progress varies greatly, meaning that each student quickly finds themselves working on an exercise suited to their level but different from that of their classmates. What starts as a single demonstration soon becomes multiple demonstrations, until I end up spending the entire session juggling explanations and answering questions.

I remember how much the issue of time management used to preoccupy me. Even though I dedicated hours to these demonstrations, I still felt it wasn't enough. I couldn't provide personalised guidance to students exactly when they needed it—and if there's one crucial aspect of welding training, it's being able to assist the student at the precise moment they need support. Frustrated, one day I decided to try something different at the start of a new course: I handed out a hand-drawn sheet to each student, illustrating the exercise, the number of beads, and the sequence in which the welds had to be completed. This sheet (somewhat resembling a recipe card) also included key details such as workpiece dimensions, electrode type, amperage, polarity, and a brief description of the techniques required to complete the task.

That day, I stayed out of the welding booths, watching how the students interpreted my instructions—and, for the first time, I had all the time in the world to answer any doubts they had. At the end of the session, someone asked: "And where's the sheet for the next exercise?"

Seeing how effective this new system was, I spent a few minutes each day preparing an exercise sheet for the following session. In reality, these sheets came directly from my own personal notes—since the very beginning of my career, I had made a habit of writing down what I learned every day, paying special attention to recording the tricks, techniques, and parameters that had helped me pass my own welding qualification tests. This is something I always recommend to my students. I tell them: the best and most personalised welding practice book is the one you write yourself, based on your own experience, after proving first-hand that the methods work.

Years later, I had compiled enough exercise sheets to support welding certification courses and specialist training programmes currently offered by the Servicio Público de Empleo Estatal (SEPE). Meanwhile, I continued filling my notebook with new knowledge—from my own certification experiences to everything I had learned from training welders to pass their own qualification tests. If you are now holding this book in your hands, it is because one day my wife, Mayte, said to me: "Why don't you try getting all those exercise sheets published?"

I want to make one thing very clear: I do not claim that everything in this book is my own original work or invention. Quite the opposite—I have been fortunate enough to meet many professionals who have shared their knowledge with me, and this book is, in part, a tribute to them. As they taught me, knowledge belongs to no one; once we gain access to it, we become its messengers, and it is our duty to pass it on to help those who are just starting out.

This book is not simply about teaching you how to weld. Its true purpose is to accompany you on your journey, regardless of your level or experience. But allow me to tell you something: everything in these pages will be useless if the most important element is missing: you. This manual has no soul without your effort, dedication, and willingness to try again and again. Just like fire shapes metal, your determination will shape your learning. Remember: the knowledge you gain is not just a tool to achieve your goals—it is also a personal transformation that will help you grow both as a professional and as a human being. This book is your companion, but the true teacher is within you.

Finally, I would like to express my deepest gratitude to all those who have placed their trust in this book since its first edition in 2010. This updated version contains everything I have learned since then. As I mentioned earlier, after everything I have experienced since that first edition, I am no longer the same person who wrote that book. And so, with all humility, if you are feeling that insatiable hunger for knowledge that unites all welders, I hope you will find in these pages what my students and I spent so long searching for.

The author

Carlos Alonso Marcos
www.linkedin.com/in/carlos-alonso-marcos-440912b0/

"With a true master, the student learns how to learn—not how to memorise and obey. The company of the noble does not shape, it liberates."

Nisargadatta

About the author

Carlos Alonso Marcos is a welding specialist with over 30 years of experience in welding engineering, technical training, and consultancy in joining processes. Throughout his career, he has worked in high-demand sectors such as aerospace, naval, space, energy, and transportation, specialising in high-pressure pipe welding, structural welding, and advanced materials.

His expertise covers both practical industrial applications and compliance with international welding quality standards, including ISO 3834, EN 1090, and aerospace welding requirements. He holds key roles as an LRQA-certified welding coordinator, forensic welding consultant, and R&D technician in advanced welding applications.

He is the author of 16 technical-practical books, widely used in vocational training and higher education across the Spanish-speaking world. His works include manuals on TIG, MIG/MAG, shielded metal arc welding (SMAW), oxy-fuel welding, and technical drawing interpretation, with a strong focus on welder certification and welding process optimisation. His latest work, *Welding Practice Manual*, is set to be published by Taylor & Francis in 2025.

In academia, Carlos is an instructor at the National Reference Centre for Electricity, Electronics, and Aeronautics in Leganés and a lecturer for the Master's in Industrial Engineering programme at the University of Alcalá, where he introduces future engineers to welding fundamentals. He has delivered over 18,000 hours of training, designing specialised welding programmes and coaching welders for industry certification across various sectors.

He is also one of the founders of the "Facultad de Soldadura" educational platform, a non-profit initiative that provides practical resources for welders and industry professionals. His current work focuses on welding process optimisation, welder certification, and knowledge transfer to the industry, ensuring the highest quality and safety standards in welded manufacturing.

Carlos Alonso Marcos—Welding specialist, trainer, and technical author

Foreword

Welding, as the art and science of joining materials, has been a driving force in the evolution of our civilization. From the rudimentary forges of antiquity to the complex industrial processes of today, the ability to join metals and other types of materials has enabled the creation of structures, machines, and tools that have transformed our world. This book is intended to be a comprehensive and practical guide, first and foremost, for those who wish to enter this fascinating field, or for those who seek to perfect their skills.

Welding is much more than simply joining metal alloys. It requires a thorough understanding of the materials, the physical and chemical processes involved, and the specific techniques for performing each type of welding. A good welder must be a craftsman, a scientist, and a technician, able to combine theoretical knowledge with manual dexterity. This book is designed to provide that combination of knowledge, bringing together both the fundamentals and the practice necessary to master the most common welding processes: SMAW, TIG, MIG/MAG, and oxy-fuel.

The book begins with SMAW, a versatile and widely used process. Here, it explores basic fundamentals to advanced techniques, including electrode selection, current and polarity types, and joint design. Special attention is given to good shop practices, a crucial aspect of ensuring safe and quality work. The coated electrode practices for carbon steel are presented in a progressive manner, from the making of the first beads to approval tests, covering a wide range of positions and joint types. A section on the welding of stainless steel with coated electrodes is also included, together with an analysis of the most common defects that can occur in this process.

The text then moves on to TIG welding, which is prized for its precision and quality. This chapter covers the history of TIG, gun/torch components, electrode and shielding gas selection, along with welding operating parameters. TIG practices for carbon steel, austenitic stainless steel, aluminium alloys, and cast iron are presented in detail, with special emphasis on approval testing. Significant space is devoted to the particularities of aluminium and its welding, including the use of alternating current, the role of thermal conductivity, the importance of cleanliness and the use of noble gases, as well as temperature control in the heat-affected zone (HAZ). The effects of pulse frequency variation and wave balance are also explored, with considerations on the choice of the most suitable filler metal for each practical situation.

The third process explored is MIG/MAG welding, an efficient and versatile method for joining metals. This chapter covers the basics of the process, equipment components, modes of metal transfer to the molten pool, shielding gases, filler wires, and welding operating parameters. Attention is given to ergonomics and safety in execution, as well as environmental factors that may affect the end result. MAG practices for carbon steel, stainless steel, and aluminium are presented in a progressive manner, including welding with flux cored wires in their different types.

Finally, the manual addresses oxy-fuel welding, a process that has a long history but still has important applications in a variety of fields. This chapter covers the basics of oxy-fuel welding, the gases used, the necessary equipment, and torch flame regulation. Oxy-fuel welding practices with carbon steel are presented in detail, covering different positions and joint types.

Throughout this manual, the importance of good shop practices, safety, and quality control is emphasised. Numerous examples, tables, and figures are included to facilitate the understanding of concepts and the execution of practices. Each chapter concludes with an analysis of common defects in the corresponding welding process and accurate recommendations on how to avoid them.

This manual is the result of years of experience and dedication to welding by its author. But, above all, it is the result of his enormous enthusiasm and love for teaching, which has led him to value as essential the proper training that welders must receive, when it is intended, above all, to weld with solvency and quality. I hope it will be a valuable tool for all those who wish to learn systematically and improve their skills in this exciting field. Welding is a craft that requires practice and dedication, but with proper guidance and dedication, anyone can master it. We invite you to immerse yourself in this manual, to practice diligently and to discover the creative and technical potential that welding offers.

In short, a work that is destined not only to be read with care, but also to be worked on and used "on the job", making the most of the abundant information it contains and taking advantage of the valuable advice that the author gives to all of us who are apprentices in welding.

Juan C. Suárez-Bermejo
Professor of Welding and NDT
Department of Materials Science
Universidad Politécnica de Madrid (UPM)

Acknowledgements

My deepest gratitude to everyone who has contributed to this book—whether with a word, a critique, a gesture, or a source of inspiration. It is a privilege to witness such beauty.

To Darío, thank you for your courage and authenticity, for teaching me that all one needs to do to see is simply open their eyes.

To Mayte, my life companion, thank you for every day, every gesture, and every sacrifice. In your imperfections, I find the charm of what is real; in your virtues, the inspiration to become better. We have walked together along paths that few dare to tread. We have seen each other at our best and our worst, and through it all, we have never stopped choosing each other.

A love that does not rely on youth or novelty is the certainty that we have found a home in one another.

To Lola, Momo, Zoe, Wanda, Cora, Nova, Runa, and Hayat.

Introduction to quotes

In the world of welding, we do not only build structures—we also weave stories. Every arc, every joint, and every effort reflects fundamental values: respect, teamwork, wisdom, and personal growth.

These quotes are not just words; they are beacons that guide our craft and our lives. Each chapter begins with one, serving as a reminder that welding is not just a technique but also an art filled with humanity and purpose.

DOI: 10.1201/9781003422488-1

1 Stick Welding (SMAW)

RESPECT—THE ART OF KINTSUGI

An ancient Japanese tradition tells of how, when a ceramic vessel breaks, it is repaired with gold, highlighting the cracks instead of concealing them. In this way, the object not only regains its usefulness but is also transformed into something unique and beautiful. This practice teaches us that scars are not a source of shame, but rather marks of resilience and learning.

Respect the scars of your journey; they are what make you strong and worthy.

INTRODUCTION TO SHIELDED METAL ARC WELDING (SMAW)

If you give me your full attention while reading this introduction to shielded metal arc welding (SMAW), I promise to focus only on the essentials. Understanding a few key concepts will make our practical work much easier and provide useful tools when the exercises become more challenging.

Shall we begin?

FUNDAMENTALS OF SMAW

The principle behind SMAW welding is the establishment of an electric current when the electrode holder—connected to a power source (welding machine)—comes into contact with the base metal, which is also connected to the same power source through the ground clamp. The current, due to the Joule effect, heats the contact area between both, primarily the tip of the electrode. The amperage selector on the welding machine controls the current flowing through the closed circuit.

By striking the electrode against the metal—like lighting a match—the spark initiates what we call the electric arc. The temperature generated in this zone can reach approximately 6,000°C (well above the melting point of most metals), causing both the tip of the electrode and the joint (the area to be welded) to transition from a solid to a liquid state. As the electrode is consumed, the welder moves the molten weld pool along the joint. The portion of the weld pool that is no longer under the arc gradually solidifies, forming what we call the weld bead.

Throughout the welding process, the electrode melts in small droplets, which are deposited into the molten pool. When only about five centimetres of the electrode remain, it is replaced, and the bead continues until the joint is fully welded. Part of the gas released from the electrode coating (CO_2) protects the molten pool from contact with atmospheric oxygen and nitrogen. Inside the electrode's coating lies the core wire, which is usually of the same composition as the base metal and varies in length and diameter. The coating type also varies depending on the application, as will be discussed later.

The electric arc is the heat source for many welding processes, providing high concentrations of heat and radiation. Due to this radiation, welders must always wear personal protective equipment (PPE), as exposure is harmful to both skin and eyes. We can describe the electric arc as a high-current discharge that transfers energy from the electrode to the workpiece through the gases produced by the electrode flux. However, under normal conditions, gases are almost perfect insulators.

So how can a gas conduct electricity?

DOI: 10.1201/9781003422488-2

3

When the electrode strikes the workpiece, the resulting heat ionises the gas, meaning that it undergoes a phase change: from gas to plasma. This transformation allows the gas to become electrically conductive instead of acting as an insulator. Once the electrode is slightly lifted from the workpiece, the ionised plasma maintains the flow of current, thereby sustaining the arc. The presence of easily ionisable materials—such as sodium and potassium—in the electrode flux facilitates this reaction.

The flow of current consists of electrons moving from the negative pole (cathode) to the positive pole (anode) of the welding power source.

The power supply of a modern welding machine includes a transformer, which converts mains electricity (high voltage and low current) into a form suitable for welding (low voltage and high current), and a rectifier, which converts alternating current (AC) into direct current (DC).

HISTORICAL BACKGROUND OF SMAW

You should know that welding is a relatively young science. The first coated electrodes were manufactured in 1912. A few years earlier, bare electrodes—without any coating—were already in use, but they produced low-quality welds and were not widely adopted. At the time, oxyacetylene welding offered better performance and remained the preferred method. However, once coated electrodes began to be mass-produced, their price dropped, and they quickly gained popularity in industry. The ingnition sequence of the arc is illustrated in Figure 1.1 and the basic electrical connections in SMAW are shown in Figure 1.2.

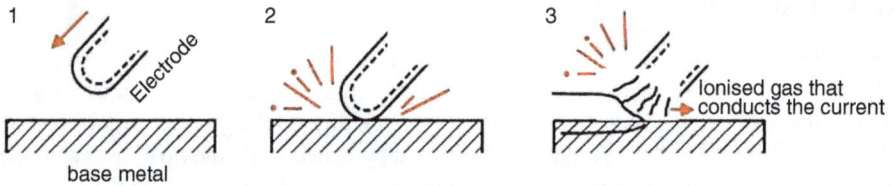

FIGURE 1.1 Sequence of arc ignition in shielded metal arc welding. The electrode touches the metal base, a spark forms as it is lifted, and the ionised gas enables current conduction.

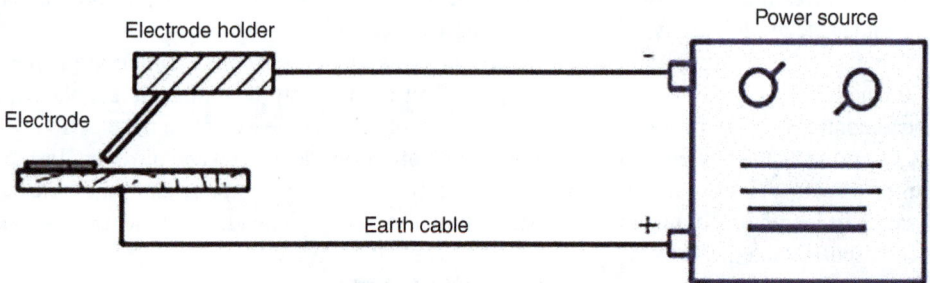

FIGURE 1.2 Diagram of the connection in shielded metal arc welding. The electrode is held in a holder connected to the power source's negative terminal, while the ground cable is attached to the positive terminal.

Today, they are not much different from those early versions. Manufacturers have significantly improved the core composition and flux coating, but essentially, it remains a technology inherited from the previous century, and its usage techniques have remained largely the same. Why is it still used? Although research has led to the development of more modern manual welding processes, such as MIG/MAG and TIG, SMAW remains widely used because of its many advantages:

- It is suitable for both indoor and outdoor welding (as long as there is no rain or strong wind).
- The equipment is relatively inexpensive and has become compact and easy to use compared to other manual welding processes.
- Thanks to the shielding effect of the flux coating, no additional shielding gases or auxiliary systems are required. Advances in electrode technology allow SMAW to be used for high-level applications, such as pressure vessels and high-pressure pipelines.
- It is suitable for a wide range of metals and alloys.

CHOOSING AN ELECTRODE

As we already know, an electrode is a metal rod specifically designed to conduct electricity and serve as filler material in manual arc welding.

During practical exercises, we will use different types and diameters of electrodes. In professional welding, you will need to select them yourself, so it is important to understand the following:

- The type of electrode depends on the welding position and the mechanical service requirements of the weld. As we will see, welding can be performed in multiple positions—some easier than others—but all must meet the same quality standards. The required mechanical properties of the weld should be specified beforehand, allowing the welder to select the appropriate electrode (for example, based on load-bearing capacity).
- The diameter of the electrode depends on the thickness of the metal and the type of weld. Electrodes are available in the following standard diameters: 1.5mm, 2mm, 2.5mm, 3.25mm, 4mm, 5mm, and 6mm. The most commonly used sizes are 2.5mm, 3.25mm, and 4mm.
  o Smaller electrodes are used for thin materials and light welding (e.g., locksmithing).
  o Larger electrodes are used for medium and heavy-duty welding (e.g., structural and boiler-making applications).

The following table shows recommended electrode diameters and amperage settings for different plate thicknesses in carbon steel:

The length of electrodes is standardised: 300mm, 350mm, 450mm, and 600mm.

What does this mean?

The diameter, length, and other characteristics of electrodes are regulated by European standards (EN) and American Welding Society (AWS) standards. Table 1.1 summarises the relationship between electrode diameter, plate thickness, and welding current.

TABLE 1.1

Recommended amperage according to plate thickness and coated electrode diameter

Plate thickness (mm)	Electrode diameter (mm)	Amperage (A)
2 to 4	2.5 to 3.25	60 to 120
4 to 6	3.25 to 4.0	100 to 200

TYPES OF ELECTRODES

A coated electrode consists of two parts:

1. The core (also called the rod or wire core)—a cylindrical metal rod that usually has the same composition as the base metal being welded.
   o   Example: A carbon steel electrode for welding carbon steel.
   o   A stainless steel electrode for welding stainless steel, and so on.
2. The flux coating, which has the following functions:
   o   Stabilises the electric arc.
   o   Shields the molten weld pool from atmospheric contamination.
   o   Reduces impurities in the weld.
   o   Controls cooling rate, preventing rapid solidification (never cool a weld bead quickly!).
   o   Introduces alloying elements that enhance the mechanical properties of the weld, such as strength and hardness.

The most commonly used electrodes are classified based on the composition of their flux coating:

BASIC ELECTRODES

- Contain calcium carbonate.
- Used for critical structural welds, low-alloy steels, and carbon steels.
- Suitable for all welding positions.
- Must be stored in dry conditions—they absorb moisture from the air, which can lead to weld defects.
- If in doubt, check the manufacturer's drying and storage recommendations on the packaging.
- The slag is typically dark brown or glossy black.
- Must be used with a very short arc length (holding the electrode too far from the work-piece will extinguish the arc).
- Compatible with both DC and AC welding machines.

RUTILE ELECTRODES

- Contain titanium dioxide (rutile).
- Used for general-purpose welding, where high mechanical strength is not required.

- Easier arc striking and reignition compared to basic electrodes.
- Suitable for all welding positions.
- Compatible with DC and AC welding machines.

CELLULOSIC ELECTRODES

- Contain organic materials in the flux coating.
- Generate a more intense arc, increasing heat input into the weld.
- Burn faster, making them ideal for vertical-down welding.
- Commonly used for root pass welding in large-diameter pipelines.

Each electrode has a code printed on its flux coating, following the AWS classification—which is simpler than the UNE standard.

FOR EXAMPLE:

- E6013 → Rutile electrode
- E7018 → Basic electrode
- Both are used for carbon steel welding.

Let's break down the meaning of these five-digit codes:

- First letter "E" → Indicates it is a coated electrode for manual arc welding.
- First two digits (60 or 70) → Indicate tensile strength in thousands of psi (pounds per square inch):

Code	Tensile Strength
60	62,000 psi (43.4 kg/mm²)
70	72,000 psi (50.4 kg/mm²)
80	80,000 psi (56 kg/mm²)
90	90,000 psi (63 kg/mm²)

- Third digit ("1", "2", "4") → Indicates welding position suitability:
  - o 1 → All positions
  - o 2 → Horizontal only
  - o 4 → All positions, especially vertical-down
- Last digit ("3", "8") → Specifies flux type and polarity compatibility.

0: DC(+) Cellulosic with sodium silicate
1: AC/DC(+) Cellulosic with potassium silicate
2: AC/DC(-) Rutile with sodium silicate
3: AC/DC(+ or -) Rutile with potassium silicate
4: AC/DC(+ or -) Rutile with iron powder (high deposition rate)
5: DC(-) Basic low-hydrogen with sodium silicate
6: AC/DC(+) Basic low-hydrogen with potassium silicate
7: AC/DC(-) Acid with iron powder and iron oxide (high deposition rate)
8: AC/DC(+) Basic with potassium silicate and iron powder, low-hydrogen (high deposition rate)

Therefore, we can say:

- E6013: A rutile electrode, suitable for all positions, for use with AC or DC welding machines (can be connected to both positive and negative polarity). It has an approximate tensile strength of 43 kg/mm².
- E7018: A basic electrode, suitable for all positions, for use with AC or DC welding machines (preferably connected to the positive pole, although it can also be used with negative polarity). It has an approximate tensile strength of 50 kg/mm².

For stainless steel electrodes, the AWS code is also printed for easier identification. For example, E309L-18:

- The three digits after the "E" refer to the AISI designation of the metal.
- The penultimate digit indicates position suitability:
  o "1" → Suitable for all positions
  o "2" → Suitable for horizontal position
  o "3" → Suitable for all positions, including vertical-down
- The last digit indicates the coating type, following the same classification as carbon steel electrodes.

Additionally, a letter may follow the first three digits:

- "L" → Low-carbon content (<0.03%)
- "H" → Medium-carbon content (0.04–0.08%)
- No letter → High-carbon content (>0.08%)

Therefore:

- E309L-18: A low-carbon basic electrode, suitable for all positions except vertical-down, for use with AC or DC welding machines (recommended to be connected to positive polarity, but can also be used with negative polarity). It has an approximate tensile strength of 49 kg/mm².

CURRENT TYPES AND POLARITY

The type of current used in welding is not always the same. We distinguish between direct current (DC) and alternating current (AC) welding machines.

Direct current (DC): These are the welding machines we will use for most of our practical exercises, except for TIG welding of aluminium, which requires AC welding machines.

Without going into excessive detail, DC welding allows the electrode to be connected to either positive or negative polarity, depending on the manufacturer's recommendations for the specific electrode type. The configurations are as follows:

- Straight polarity (DC-): Electrode to negative (-) and workpiece to positive (+).
- Reverse polarity (DC+): Electrode to positive (+) and workpiece to negative (-).

In DC welding machines, electric current always flows from the negative pole to the positive pole. This has a crucial consequence: the component connected to the positive pole receives more heat. The heat distribution is approximately:

- 70% of the heat is concentrated in the positive pole.
- 30% is concentrated in the negative pole.

Additionally, the arc is more stable than in AC welding machines.

Alternating current (AC): Most modern welding machines feature a current selector to switch between DC and AC. However, basic welding machines (unless they are older transformer-based models) generally provide DC only.

For SMAW and MIG/MAG, we will use DC, since it provides a more stable arc and allows better heat distribution control through polarity selection.

However, for TIG welding of aluminium, AC is necessary—a topic that we will discuss in more detail later.

JOINT DESIGN

When welding, penetration (the depth at which fusion occurs) is limited. In some cases, full penetration cannot be achieved unless the material is thin.

A general rule is that if the workpiece thickness exceeds 3–4mm, the edges should be bevelled to allow the arc heat to reach the opposite face of the joint, ensuring proper fusion of the edges and forming what is known as the root pass. The geometry of these bevels is what we call joint design.

The specific joint design depends on:

- The thickness of the workpiece.
- Whether the joint can be welded from one side only (e.g., small-diameter pipes) or from both sides.
- The welding process used.

Each scenario requires a different joint configuration to ensure quality welds (Figure 1.3).

Simplified overview of welding joint types (Include symbols alongside the representation of each joint type)

a. **Square groove joint.** If the workpieces have a thickness of no more than 3–4mm, they can be butt welded (placed edge to edge, with or without a gap between them, ensuring they are level with each other) (see Figure 1.4).

b. **Single V-groove joint:** The edges of the workpieces are bevelled when their thickness exceeds **3–4mm**, as we will see in the practical exercises. The bevel angle varies depending on the material thickness, the type of metal, and the welding process used. This type of edge preparation is applied to **moderately thick workpieces** that can only be welded from one side.

   A **small land** (a flat edge left at the root of the joint) is usually maintained, and a **root gap** (space between the workpieces) is also left to facilitate fusion (see Figure 1.5).

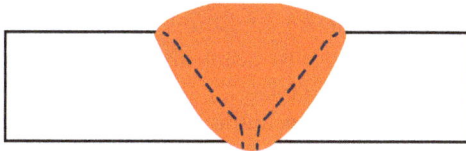

FIGURE 1.3   Cross-section of a V-groove butt weld. The filler metal (in orange) fully fills the bevel, ensuring fusion with the base metal.

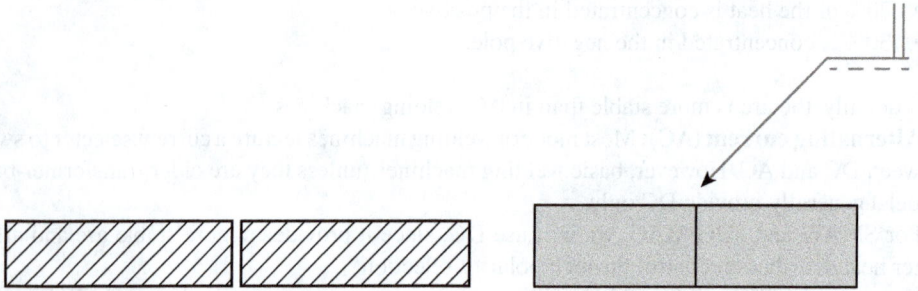

FIGURE 1.4   Two metal pieces in cross-section, prepared for a square butt joint.

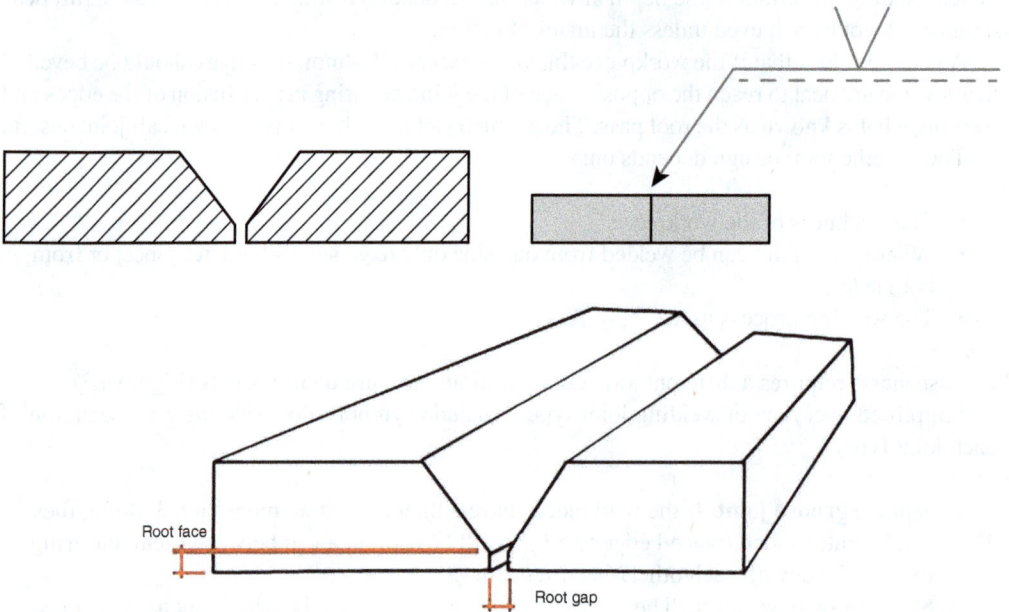

FIGURE 1.5   Preparation of a V-groove butt joint with root face and root gap. The gap between the pieces and the un-bevelled material thickness at the root are shown.

c. **Double V-groove joint.** For workpieces thicker than 10–12mm, it is advisable to use symmetrical bevels, as they significantly reduce the stresses and distortions caused by welding compared to a single V-groove. This preparation requires access to both sides of the joint to ensure proper execution of the weld (see Figure 1.6).

d. **U-groove joint.** For thick materials that can only be welded from one side, a J-bevel is recommended, forming a U-groove joint.

   This type of preparation requires machining the edge to achieve the proper shape (see Figure 1.7).

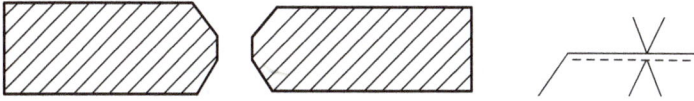

FIGURE 1.6 Cross-section of a V-groove butt joint preparation. Both pieces have a symmetrical bevel with a gap between them to allow weld penetration.

FIGURE 1.7 Cross-section of a weld with lack of penetration. A gap is visible at the root of the joint, indicating a separation between the pieces for the root pass.

FIGURE 1.8 Butt joint of two pieces.

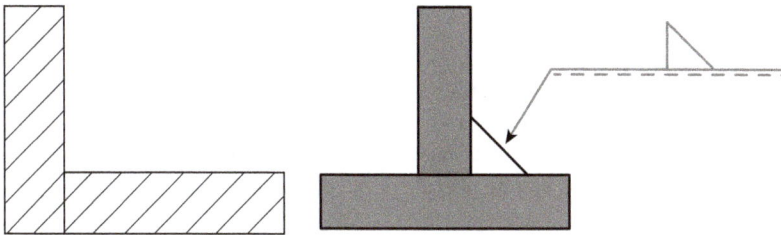

FIGURE 1.9 Fillet joint of two pieces.

These joint designs are applied to the following types of weld joints:

1. **Butt joint** (see Figure 1.8)
2. **Fillet joint** (see Figure 1.9)
3. **Lap joint**

The resulting weld bead appearance obtained wih 7016 basic electrodes is presented in Figure 1.10.

TECHNOLOGICAL TOOLS

Arc force and arc dynamics (or inductance) are key parameters in welding machines, especially in processes such as SMAW. Let's examine the difference:

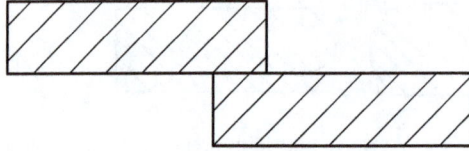

FIGURE 1.10   Lap joint of two pieces.

i.   **Arc force**

- **What it is:** Controls the intensity of the current when the electrode approaches the base material.
- **Effect:**

    o **Increasing it:** Provides more current when striking the electrode, preventing it from sticking and improving penetration. The arc becomes more aggressive and focused.
    o **Decreasing it:** The arc becomes softer and less penetrating, but there is a higher risk of the electrode sticking if accidental contact occurs.

ii. **Arc dynamics or inductance**

- **What it is:** Controls the speed at which the current increases and decreases during the arc cycle.
- **Effect:**
    o **Increasing it:** Produces a smoother arc and a more fluid weld pool, ideal for short-circuit welding, reducing spatter and improving the bead's appearance.
    o **Decreasing it:** The arc becomes more abrupt, with more spatter and a more controlled weld pool, which can be useful for deeper penetration.

What happens when both are adjusted together?
- **High arc force + high inductance:** Greater control in difficult welds, with increased penetration and a more fluid weld pool—ideal for challenging positions.
- **Low arc force + low inductance:** A more stable and less aggressive arc, suitable for finishing beads and achieving cleaner welds on thin materials.

All of this is theoretical, but throughout the chapter on electrodes, we will explore practical examples of how to apply these tools in real welding situations. Figure 1.11 ilustrates the clamping arrangemente with tack-welded auxiliary plates used during the PF welding process.

PROCESS APPLICATIONS

As I promised at the beginning of the theory chapter, I will now conclude. SMAW is not only used for welding carbon steel but also for:

- Alloy steels
- Stainless steels
- Cast iron

FIGURE 1.11 PF weld in carbon steel with 7016 basic electrode, clamped with tack-welded auxiliary plates.

- Other metals, such as bronze alloys
- In some cases, such as aluminium alloys, the quality achieved with electrodes is not ideal, and it is preferable to use MIG or TIG welding.

Currently, it is not possible to weld low-melting-point metals, such as lead or tin, with shielded metal arc welding (Figure 1.12). Likewise, it is not suitable for highly oxygen-sensitive metals,

FIGURE 1.12   Welding of Carbon Steel Pipe with Electrode in 5G Position

FIGURE 1.13   Construction of an aluminium boat in a workshop. The structure is in the welding and assembly process, with the author working on a scaffold next to the vessel.

Image courtesy of ALUTECH MARINE

such as titanium, since the gases released by the electrode coating are insufficient for proper protection. Figure 1.13 illustates the welding and assembly process of an aluminium boat.

RECOMMENDATIONS

The most suitable electrodes for steels with impurities or high carbon content are basic electrodes, followed by rutile electrodes. Cellulosic electrodes are the most sensitive to these impurities.

For very thin materials, rutile electrodes work well because they allow easy stopping and restarting, providing better control over fusion.

In edge preparations with larger-than-planned gaps, electrodes that cool the weld pool quickly and offer high toughness are needed—such as basic and rutile electrodes.

"Every great welder started with doubts and mistakes. What matters is not how many times you fail, but what you learn from each attempt. This manual is not only intended to guide you technically, but also to remind you that every weld bead is an opportunity to improve yourself and prove what you are capable of. Be patient with yourself; mastery is forged with time and perseverance, just like a good weld."

VERY IMPORTANT: BEST PRACTICES IN THE WELDING WORKSHOP

- **Slag protects the weld:** The electrode coating turns into slag, which shields the weld while it is still hot. Do not remove it before it cools down, as this may compromise the weld quality.

- **Always wear eye protection when removing slag:** When chipping off slag with a chipping hammer, always wear safety goggles. Slag particles can fly off and cause injury.
- **Make use of waiting time:** While the slag cools, take advantage of the time to do something useful, such as removing spatter (those small metal droplets that form around the weld). This will significantly improve the final appearance of your work.
- **Use tools, not your hands, to handle hot pieces:** The welded pieces will be extremely hot. Use pliers or tongs to manipulate them. Even if you wear gloves, remember that heat can be intense, and you could burn yourself.
- **Ask your instructor if in doubt:** If you are unsure how to do something, especially when using tools such as grinders, saws, or sanders, ask for help. These tools require respect but not fear. Always use them with caution.
- **Never look at the arc without proper protection:** Never look directly at the arc without a welding helmet. The intense light can seriously damage your eyes.
- **Never work without gloves or in short sleeves:** Always wear the recommended protective gear: apron, sleeves, gaiters, mask, etc.
- **Use the electrode to the very end:** An electrode is considered used up when less than 5cm remains. Use it completely before discarding it.
- **Strike the arc correctly:** To ignite the electrode, think of lighting a match. Scratch it with a quick and precise movement until you see the arc light up, then bring it to the starting point of the weld.
- **Practice bead overlaps:** In every exercise, practice how to join weld beads. This is a fundamental skill that will help improve your welds.
- **Be organized and take care of materials:** Take care of your equipment and materials, and help clean the workspace at the end of the session.
- **Read the instructions carefully before starting:** Review the guidelines for each practice carefully before beginning. Once you understand what to do, observe the piece and trust your intuition.
- **Believe in yourself:** If others can do it, so can you. Welding is about understanding the technique and practicing until you master it.
- **Attitude is everything:** To succeed in practice, you must believe that you can achieve it. If you don't trust yourself, it will be much harder.
- **Work as a team:** Do your training exercises with a partner. Two heads think better than one, and it is easier to detect someone else's mistakes than your own.
- **Train your hand movements:** While welding, you must control the distance, angle, and travel speed of the electrode. Keep the arc light as low as possible to clearly see the bead forming. If you maintain the correct distance, it will be easier to move forward at a consistent speed.

SHIELDED METAL ARC WELDING

PRACTICE 1. FIRST WELD BEADS + OVERLAY WELDING IN HORIZONTAL POSITION. PA(1G)

- **Base material:** Carbon steel plate 4″ x 4″ x 1/8″.
- **Electrodes to be used:** Ø 3/32″ x 14″ rutile electrode 6013. Direct polarity.
- **Number of weld beads:**
  Four weld beads deposited using a straight movement.
  Two overlay beads applied with a half-moon motion.
- **Welding current:**
  60–65 amps for weld beads with straight movement.
  70–75 amps for overlay weld beads.

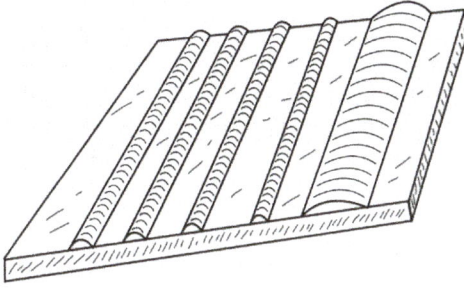

FIGURE 1.14   Metal plate with multiple weld beads arranged in parallel. A progression in weld quality is visible, from irregular beads to a smooth, uniform finish.

Welcome to your first welding experience! Let's begin with two simple exercises (Figure 1.14).

In this Practice No. 1, we will execute straight weld beads, which will later serve as guides for what we will call "overlay" or "grooming" beads.

In the next practice (No. 2), we will learn how to overlap weld beads.

Let's be clear from the start: Although these exercises are simple, they form the foundation for all the subsequent welding practices. Therefore, **WE MUST MASTER THEM BEFORE MOVING ON TO PRACTICE No. 3.**

MATERIAL PREPARATION

Before starting, it is highly recommended to mark straight lines on the base material (the metal plate) using a scribing tool and then reinforce them with a small grinder and cutting disc or a centre punch to create clearly visible references. These lines will be our only reference to maintain a straight bead during welding.

EXECUTING THE STRAIGHT BEADS

Once the marking is complete, we can begin:

- Start by striking the arc on one of the centre-punched lines, scraping the electrode tip against the plate like a match until the arc ignites and becomes visible through the dark glass of the welding helmet.
- Once the arc is established, move the shielded metal arc electrode forward along the line.

We must focus on maintaining three essential conditions:

1. Hold the electrode at a slight angle, about 70°–80° in the direction of travel.
2. Maintain a constant distance of 2–3mm between the electrode tip and the base metal.
3. Advance at a steady speed while ensuring the previous two conditions are met. This means keeping the correct angle while maintaining a consistent electrode-to-metal distance as the electrode gets shorter while it burns.

   Allow the molten weld pool (the liquid metal puddle beneath the electrode) to expand slightly before advancing.

   As a reference, the diameter of the molten pool should be 1.5 to 2 times the diameter of the electrode.

Complete all straight weld beads first, then remove the slag using a chipping hammer, and finish with a thorough wire brushing to clean the surface of the plate (Remember to wear your personal protective equipment).

Now, let's fill the gaps between the straight beads.

When we say "overlay" or "grooming", we refer to weld beads that are significantly wider than the electrode diameter. This is achieved by moving the electrode laterally in a U or Z motion (see sketch).

The idea is to move across the two straight beads, filling the space between them with molten metal from the fusion of both the electrode rod and the surface of the plate, heated by the welding arc.

I.   Important:

While performing the U or Z motion:

- **Pause for a moment at each side** of the defined space for about **two seconds** before moving to the opposite side.
- **Repeat this pattern** until the electrode is fully consumed.
- Holding the electrode **steady at each side** ensures **a uniform weld bead**.
- **Overlap the lateral movements**, ensuring the space is entirely filled with no empty areas.

II.   Techniques and additional tips:

1. **Practice without striking the arc.** Get used to the hand position and movements before turning on the welder.
2. **Monitor the molten pool.** Closely watching the molten metal puddle helps adjust travel speed and electrode angle.
3. **Control your breathing.** Maintaining steady diaphragm breathing can improve fluid and controlled movements.
4. **Take your own notes.** Keeping personal records of challenges and solutions is a valuable habit for progress.

Even if you have previous welding experience, follow these exercises precisely.

Bad habits are harder to unlearn than learning from scratch!

Good luck!

SHIELDED METAL ARC WELDING

PRACTICE 2. FIRST WELD BEADS: LEARNING TO MAKE OVERLAPS

- **Base material:** Carbon steel plate 4" x 4" x 1/8".
- **Electrodes to be used:** Ø 3/32" x 14" rutile electrode 6013. Direct polarity.
- **Number of weld beads:** Eight weld beads, all executed with a straight movement.
- **Welding current:** Between 55 and 70 amps.

This simple operation is fundamental. It is essential that every time an electrode is consumed, the welder is able to overlap a weld bead seamlessly, so that the junction is barely noticeable. This practice consists of executing straight weld beads while applying the following technique to learn how to properly overlap welds.

I.   Instructions:

1. At the end of the weld bead, a small "ramp" forms, which will serve as a guide for achieving a seamless overlap.

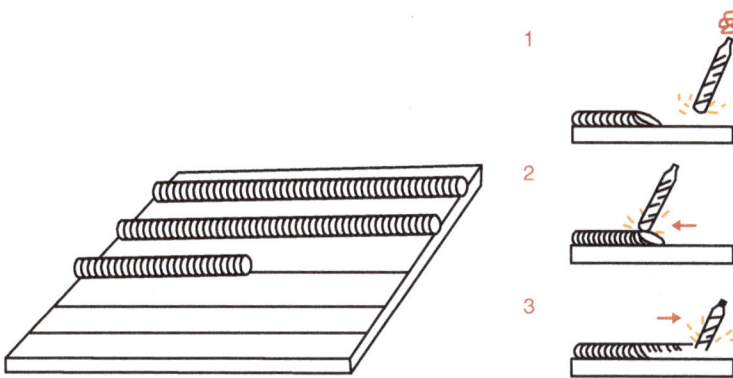

FIGURE 1.15   Metal plate with parallel weld beads. On the right, three diagrams show how to make a weld splice.

2. Insert a new electrode into the holder and start the arc a couple of centimetres ahead of the ramp (see Figure 1.15).
3. Once the arc is established, slowly move backward towards the ramp while keeping the arc active, and climb up the ramp until you reach its highest point.
4. At that exact point, stop for a moment before advancing in the opposite direction at a steady speed, maintaining an electrode inclination of approximately 70°–80° and a gap of 2–3mm between the electrode tip and the workpiece (see Sketch No. 2 and No. 3).

II.   Weld overlap evaluation:

✓ If executed correctly, the ramped area will have fully fused into the new weld bead, making the transition visually seamless.
✓ If the overlap is excessively raised, it means the electrode has been moved beyond the highest point of the ramp, overlapping the previous weld bead.
✓ If there is a visible depression at the overlap, it indicates the electrode did not reach the highest point of the ramp before advancing.
✓ The approach movement must be slow and precise to ensure the electrode is positioned exactly at the correct point before proceeding.

SHIELDED METAL ARC WELDING

PRACTICE 3. FILLET WELD IN HORIZONTAL POSITION. PA(1F)

- **Base material:** Carbon steel plate. Two pieces of 1.6″ x 4″ x 1/8″ tack-welded at 90°.
- **Electrodes to be used:** Ø 3/32″ x 14″ rutile electrode 6013. Direct polarity.
- **Number of weld beads:** Six weld beads, all executed with a straight motion.
- **Welding current:** 70–80 amps.

This is the first real joint welding practice. In this exercise, we will fill the gap in this V shape with straight weld beads, following the sequence shown in the diagram (Figure 1.16). Between each pass, always remove the slag, clean off the spatter (remember, those small metal droplets that ruin the appearance of a well-made weld), and brush the weld bead(s) thoroughly.

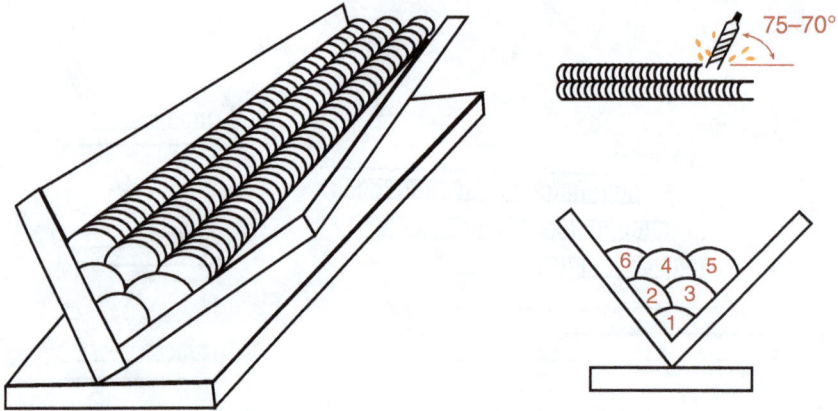

FIGURE 1.16   Welding in a V-groove joint with multiple filler passes. On the right, diagrams show the electrode inclination angle and the bead deposition sequence in the groove.

The recommendations remain unchanged:

- Maintain a 2–3mm gap between the electrode tip and the plate.
- Keep the electrode at a 70°–80° angle in the direction of travel (never tilt it laterally!).
- Advance at a constant speed.

I.   Techniques and tips:

☑ **You have two hands—use them!** Hold the electrode holder firmly but without excessive force, and keep your arm as relaxed as possible. If your arm is too stiff, you will struggle against yourself and lose precise control over the electrode movement.

☑ In this exercise, **slow down the electrode's travel speed** compared to previous practices. As a reference, the **weld bead width should be approximately twice the diameter of the electrode tip**.

☑ **Keep your welds straight!** Find a visual reference to maintain a **straight travel path**.

II.   Weld bead sequence:

- The second and third weld beads are deposited over the first one:
  o Bead No. 2 should partially cover Bead No. 1.
  o Bead No. 3 should overlap half of Bead No. 2.
- The fourth, fifth, and sixth beads are deposited over the previous ones:
  o Bead No. 4 should be centred along the line between Beads No. 2 and No. 3.
  o Beads No. 5 and No. 6 should be positioned on either side of Bead No. 4.

SHIELDED METAL ARC WELDING

PRACTICE 4. FIRST WELD BEADS IN HORIZONTAL POSITION (PC-2G)

- **Base material:** Carbon steel plate 4″ x 4″ x 1/8″.
- **Electrodes to be used:** Ø 3/32″ x 14″ rutile electrode 6013. Direct polarity.
- **Number of weld beads:** Between 15 and 20 weld beads, executed with a straight movement, starting from the lower part of the plate.
- **Welding current:** 55 to 65 amps. Adjust the intensity to a level that allows for comfortable welding (Figure 1.17).

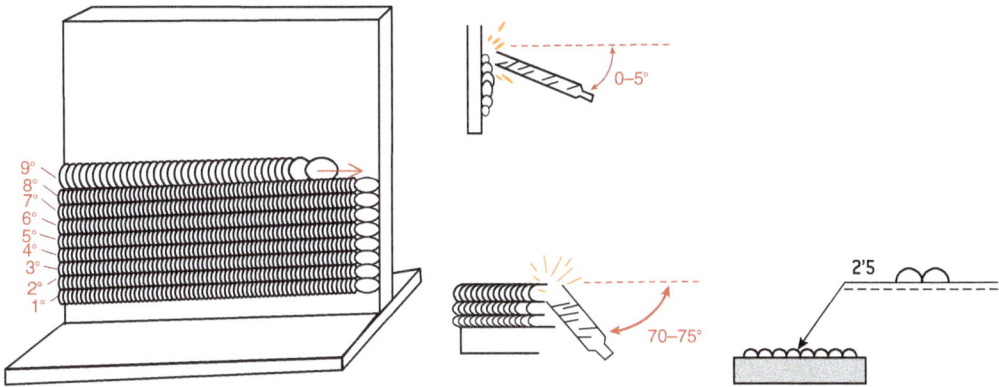

FIGURE 1.17   Vertical-up welding with a shielded metal arc electrode. On the right, diagrams show the electrode inclination angle and the travel direction in the zigzag technique.

OVERLAY WELDING: A REPAIR TECHNIQUE TO RESTORE THICKNESS

An overlay weld is a type of repair welding used to restore thickness to a part that has worn down over time. Before welding in the horizontal-vertical (PC-2G) position, where gravity works against us, it is recommended to practice in the flat position first to get familiar with the technique.

I.   Step by step:

1. **Initial positioning:**
   - Secure the plate firmly in place to prevent movement during welding.
2. **Starting the first bead:**
   - Begin at the bottom of the plate.
   - Hold the electrode at an inclination of 70°–80° relative to the joint.
   - Maintain an electrode tip-to-base metal distance of 0.08–0.12″ (2–3mm), just as in the previous practice.
   - Adjust the current intensity to ensure a smooth and comfortable welding process, achieving a bead appearance similar to the one obtained in the flat position.
3. **Electrode motion:**
   - Move in a straight and steady motion, avoiding any tilt other than the forward travel angle.
   - Ensure the electrode tip remains close to the molten pool, preventing the bead from sagging or forming "bulges".
4. **Overlapping the beads:**
   - For the second and subsequent beads, aim for the line where the previous bead ended.
   - Each new bead should overlap approximately half of the previous bead.
   - Overlap all beads properly to ensure no gaps remain between them.
5. **Inspection:**
   - After completing all the beads, inspect the weld to confirm there are no gaps and that all beads are level and properly fused.

Tips:

- ✓ **Patience and control:** Welding in the PC-2G position is challenging because gravity tends to pull the molten metal downward. Keep the electrode tip close to the molten pool and control every movement carefully.
- ✓ **Practice:** Like any technique, consistent practice is key to mastering this position.
- ✓ **Don't get discouraged if your first attempts aren't perfect!** Keep practicing and seek advice from your instructor or experienced colleagues.

Welding in the horizontal-vertical position tests both your technique and your patience. At first, it is natural to feel frustrated because gravity is not on your side. But remember, every challenge you overcome makes you stronger.

If something doesn't turn out as expected, stop, analyze, and try again. Every weld bead is a lesson, and every practice session brings you closer to mastering this position.

Welding is not just a technical skill—it is also an exercise in focus and perseverance. Stay calm, trust your effort, and keep moving forward.

SHIELDED METAL ARC WELDING

PRACTICE 5. FILLET WELD IN THE HORIZONTAL-VERTICAL POSITION (PC-2G)

- **Base material:** Carbon steel plate. Two pieces of 1.6" x 4" x 1/8", tack-welded at 90°.
- **Electrodes to be used:** Ø 3/32" x 14" rutile electrode 6013. Direct polarity.
- **Number of weld beads:** Six weld beads, all executed with a straight motion.
- **Welding current:** Approximately 70–80 amps (Figure 1.18).

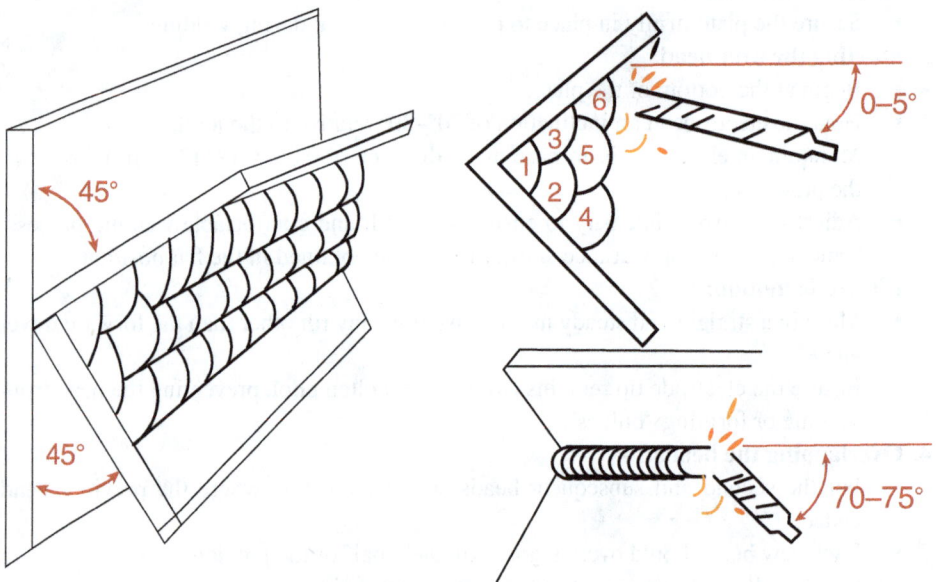

FIGURE 1.18 Welding in the ledge position with a shielded metal arc electrode. On the right, diagrams show the bead deposition sequence, electrode inclination angle, and travel direction.

The difficulty of this exercise is that you might lose focus, and before you realize it, more material accumulates on the lower plate than on the upper one. The goal is to achieve the same finish as in exercise No. 3, while applying what was learned in exercise No. 4.

⚠ **Take extra care to prevent the electrode tip from moving too far from the plate** (this will **avoid sagging metal and reduce spatter**).

⚠ Maintain all the other key parameters:
- 70°–80° inclination in the direction of travel.
- 0.08–0.12″ (2–3mm) gap between the electrode tip and the plate.
- A straight and steady travel speed, allowing the molten pool width to expand to twice the diameter of the electrode tip.

I.   Weld bead sequence:

1. **For the root bead**, aim directly at the joint.
2. Follow the order shown in the diagram for the next beads:
   - **Bead No. 2:** Aim at the first bead and weld over it as if trying to fully cover bead No. 1.
   - **Bead No. 3:** Aim at the upper line of bead No. 1 (this will also partially fuse with bead No. 2).
   - **Bead No. 4:** Aim at the lower line of bead No. 2.
   - **Bead No. 5:** Aim at the line between bead No. 2 and bead No. 3.
   - **Bead No. 6:** Aim at the upper line of bead No. 3, pausing slightly to allow proper filling of the gap.

SHIELDED METAL ARC WELDING

PRACTICE 6. LEARNING TO MAKE TACK WELDS

- **Base material:** Carbon steel plate 4″ x 4″ x 1/8″.
- **Electrodes to be used:** Ø 3/32″ x 14″ rutile electrode 6013. Direct polarity.

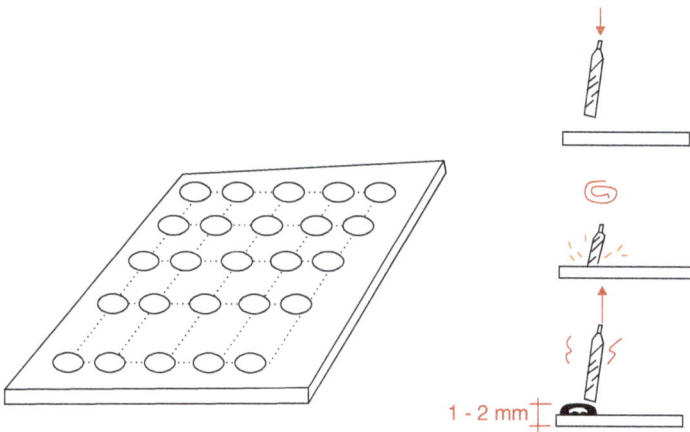

FIGURE 1.19   Metal plate with spot weld beads. On the right, diagrams illustrate the tack welding technique with a shielded metal arc electrode, including spacing distance and the arc ignition procedure.

- **Welding current:** Approximately 75–85 amps.
- **Number of tack welds:** 25 in total.
  - o From the edge of the plate, mark five points at a distance of 0.4″ (10mm) from the edge.
  - o Leave a 0.8″ (20mm) gap between each tack weld (Figure 1.19).

The so-called tack welds are used in assembly to temporarily hold a structure that will later be fully welded. Ensuring that tack welds are strong enough is a crucial task that carries great responsibility.

I.   Method

1. Mark the base metal using a centre punch at the distance specified in the introduction to this exercise. This will be your reference point for placing the tack weld.
2. Scratch the electrode against the plate with a quick and precise motion (remember, like striking a match) and position it over the marked point.
3. With the electrode held almost vertically and always maintaining a 0.08–0.12″ (2–3mm) gap between the electrode tip and the base metal, trace a circular spiral motion approximately 0.4″ (1cm) in diameter, starting from the outside and finishing at the centre.
4. Hold the arc momentarily at the centre, allowing enough filler metal to be deposited before breaking the arc.

If done correctly, a shiny, flat steel "button" of about 0.08–0.12″ (2–3mm) in height will remain on the plate, without cracks or slag inclusions.

WHAT IS WELDER QUALIFICATION?

Welder qualification is a process through which a welder is evaluated and certified to perform welding work according to specific quality and safety standards. This process ensures that the welder possesses the necessary skills and knowledge to produce high-quality welds in various positions and with different types of materials.

Certification is crucial for career advancement, especially in industries that require high precision and reliability in welding, such as construction, machinery manufacturing, and the petrochemical industry.

QUALIFICATION VALIDITY

Welder certification usually has a limited validity period.

- Generally, a welder qualification is valid for two to three years.
- After this period, the welder must renew the certification, which typically involves performing new practical tests to demonstrate continued proficiency.

COST OF CERTIFICATION

Welder qualification is not free, and costs can vary significantly depending on:

- The country where it is obtained.
- The certifying institution.
- The type of welding process and materials used in the test.

♀ Some employers may cover the cost of certification for their workers, particularly in industries where it is essential for job performance.

EDUCATIONAL REQUIREMENTS

No formal education is required to obtain a welder qualification. However, candidates are expected to have:

- Practical knowledge and welding skills.
- Experience gained through technical courses, vocational training programs, or on-the-job learning.

💡 Some certification institutions may require candidates to complete preparatory courses before attempting the qualification tests.

MOTIVATION AND SUPPORT

I understand that facing a welder qualification test can be intimidating, and doubts and insecurities may arise. However, remember:

- ☑ This is a crucial step in your professional development and a great opportunity to prove your skills.
- ☑ Every certified welder has gone through the same process.
- ☑ Trust your training and the practice you have done.

Every weld bead you make and every practice you complete is bringing you closer to your goal.

- ◇ Consistency and precision are key.
- ◇ I am here to support you every step of the way.
- ◇ If I, despite never being the smartest or the most skilled, have managed to build a career in this incredible profession, then anyone who sets their mind to it can achieve it.

CERTIFICATION PRACTICES

From this point forward, exercises labelled as "Certification practice" refer to real qualification tests commonly used in welder certification exams. Therefore, in these cases, the parameters and recommendations provided are not just for learning how to weld, but also to help pass a specific certification test.

SHIELDED METAL ARC WELDING

PRACTICE 7. HORIZONTAL FILLET WELD. PB (2F)

- **Base material:** Carbon steel plate. Two pieces of 6″ x 1.6″ x 5/16″, tack-welded at 90°.
- **Electrodes to be used:**
  - o Ø 1/8″ x 14″ rutile electrode 6013. Direct polarity.
  - o Then, repeat the exercise using a Ø 1/8″ x 14″ basic electrode 7016, reverse polarity (electrode holder to + and ground clamp to -).
- **Number of weld beads:** Three weld beads, all executed with a straight motion.
- **Welding current:** 105–120 amps.

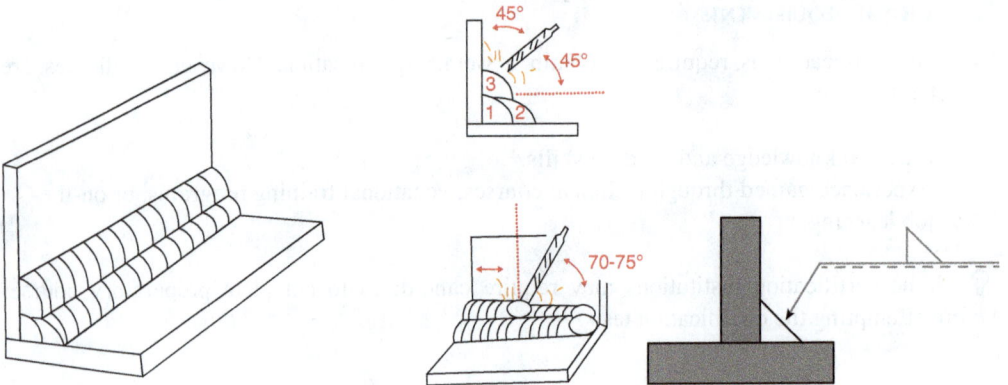

FIGURE 1.20   Welding in a fillet joint with a shielded metal arc electrode. On the right, diagrams show the bead deposition sequence, electrode inclination angle, and travel direction.

EXERCISE DESCRIPTION

Now that we have practised fillet welding, let's see how an actual welder qualification test would be conducted.

The exercise consists of executing three weld beads (in the order shown in Figure 1.20). In all three, it is essential to apply everything learned so far. Let's recap:

- Minimal electrode inclination in the direction of travel (70°–80°).
- Short arc length (between 0.08″ and 0.12″ (2–3mm) from the electrode tip to the joint).
- Straight travel with a constant speed, allowing enough time for the weld bead width to reach twice the electrode diameter.

If these parameters are not strictly followed throughout the entire process, the weld bead will be rounded instead of flat, or a groove (notch) will form between beads No. 2 and No. 3, when this overlap should be smooth and even.

I.   Technique

1. Once bead No. 1 is completed, clean it thoroughly, removing all slag and spatter.
2. Execute bead No. 2 in the same way, but aim directly at bead No. 1 as if it did not exist— it must be fully covered, not simply placed next to it.
3. For bead No. 3, aim at the upper edge of bead No. 1, ensuring it overlaps bead No. 2 and conceals its upper half.

WELDING MACHINE PARAMETER ADJUSTMENTS

To achieve optimal penetration and avoid defects:

- **Arc force:**
    - o   Set the arc force to the maximum allowed by the machine when executing the first bead (root pass).

o This will ensure greater penetration and prevent the electrode from sticking.
o A high arc force also helps stabilise the arc when the electrode approaches the base material.

- **Inductance/dynamics:**
  o Keep the setting at a medium-high level to avoid an excessively aggressive arc.
  o This will allow for a more fluid weld pool, facilitating better fusion of subsequent beads.
  o Slightly reduce the inductance when executing the final beads to control the weld pool and prevent excessive reinforcement.

I. Additional tips

✓ Adjust the current within the recommended range for this practice to balance penetration and weld pool control.
✓ Inspect each bead visually after every pass to ensure no notches or lack of fusion are present.

With these settings, you will be better prepared to achieve proper penetration and a smooth transition between beads, meeting welder qualification requirements.

Facing your **first qualification test** may bring **nervousness and high expectations**. But remember: this is just **one more step in your journey**.

You have practised, learned, and overcome challenges to get here. Qualification is not only a **technical assessment** but also a **test of your perseverance and preparation**.

◇ **Focus on every movement, every detail, and let your effort reflect in your work.**
◇ **This exercise marks the beginning of your journey towards welding excellence.**
◇ **Trust yourself and what you have learned—you are more prepared than you think!**

THE ROLE OF CSWIP WELDING INSPECTOR (CSWIP WI)

WHAT IS A CSWIP WELDING INSPECTOR?

A CSWIP welding inspector (CSWIP WI) is a certified professional responsible for ensuring that all welds in a construction project comply with specified quality and safety standards. The CSWIP WI plays a key role in welder qualification, as they are in charge of evaluating and certifying that the welds performed by candidates meet the required standards.

CSWIP WI INVOLVEMENT IN WELDER QUALIFICATION

In the welder qualification process, the CSWIP WI is responsible for supervising and assessing practical tests. This includes a visual inspection of the welds, as well as the application of non-destructive testing (NDT) and destructive testing to verify compliance with quality standards.

CSWIP WI INSPECTION OF A FILLET WELD WITH SHIELDED METAL ARC ELECTRODES

For fillet welding, whether in the horizontal position or any other, using shielded metal arc electrodes, the CSWIP WI follows a detailed protocol based on ISO 5817. This protocol includes the following steps:

1. **Visual inspection:** The CSWIP WI begins by inspecting the weld to identify surface defects such as cracks, porosity, undercut, and excessive reinforcement. The weld should be uniform, with a flat profile and no discontinuities.
2. **Dimensional measurements:** The length, width, and height of the weld bead are measured to ensure they comply with the welding procedure specifications.
3. **Evaluation according to ISO 5817:** The weld is assessed based on specific criteria in ISO 5817, which classifies welding defects into three quality levels:
   o **B** (high quality),
   o **C** (medium quality),
   o **D** (low quality).
   o For **welder qualification**, at least **quality level C** is generally required.

VISUAL INSPECTION PROTOCOL

- **Surface defects:** No visible cracks or slag inclusions are permitted. Porosity should be minimal and evenly dispersed.
- More details can be found in the defectology chapter.

DESTRUCTIVE TESTING PROTOCOL

The CSWIP WI must ensure that the weld bead has achieved approximately 0.04″ (1mm) of penetration. Penetration refers to how deep the weld fuses into the base material. While 0.04″ (1mm) may seem small, it is challenging to achieve in an 8mm thick plate, especially when both pieces are butted together with no gap.

To verify this, a fracture test is performed on the test coupon (the welded pieces used for examination). This test ensures that the penetration required for certification exams or critical welding work has been achieved.

To break the coupon:

1. Grind off the tack welds using a grinding disc.
2. Make a longitudinal cut in the root pass, centrally along its entire length.
   o The cut should be wider than deep, taking care not to reach the root or fusion line between the two plates before welding.

Next, the coupon is placed in a press or a bench vise. In the case of a vise, it must be firmly secured to ensure a safe break. The aim is to apply force on the edges of the fillet weld, attempting to close the angle until the weld fractures, allowing the CSWIP WI to inspect the internal structure (see Figures 1.21–1.22).

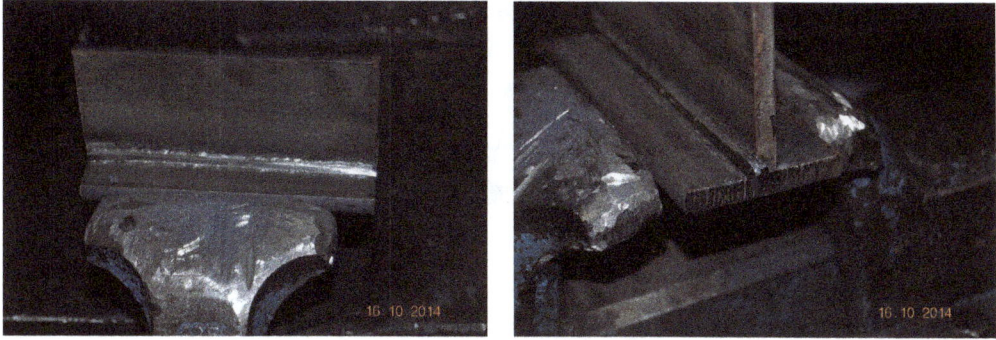

FIGURE 1.21 Unwelded fillet joint, clamped in a bench vise. The joint preparation is visible, with a small root gap.

FIGURE 1.22 Unwelded fillet joint, clamped in a bench vise. The joint preparation is visible, with a small root gap.

Next, the coupon is placed in a press or a bench vise. In the case of a vise, it must be firmly secured to ensure a safe break. The aim is to apply force on the edges of the fillet weld, attempting to close the angle until the weld fractures, allowing the CSWIP WI to inspect the internal structure (see Figure 1.23).

The measurement should be taken using a **caliper (vernier gauge)** whenever possible.

If the required 0.04″ (1mm) penetration is not achieved, the test must be repeated, adjusting the amperage settings until the correct penetration is obtained (see Figure 1.24).

The test coupon has specific **minimum dimensions**:

In a welder qualification exam, it is mandatory to perform stops with tie-ins for every bead. Two tie-ins should never be placed in the same location, as they are high-risk areas for defects. If they accumulate, they may cause serious structural issues (Figure 1.25).

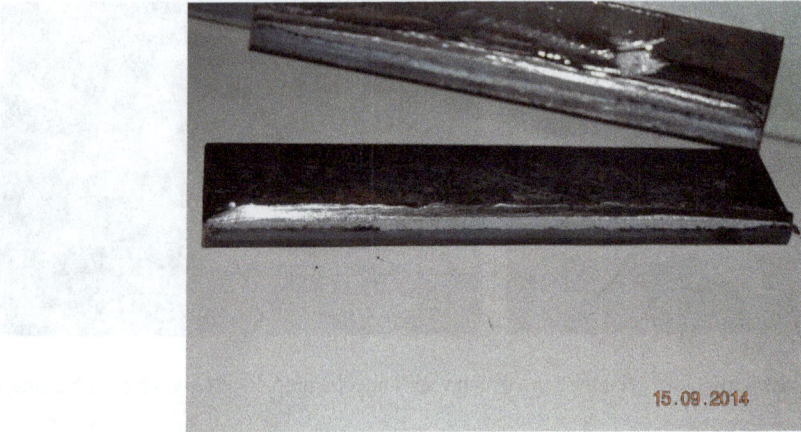

FIGURE 1.23   Fillet joint after a break test. The pieces have separated along the weld bead, revealing the extent of its penetration and fusion.

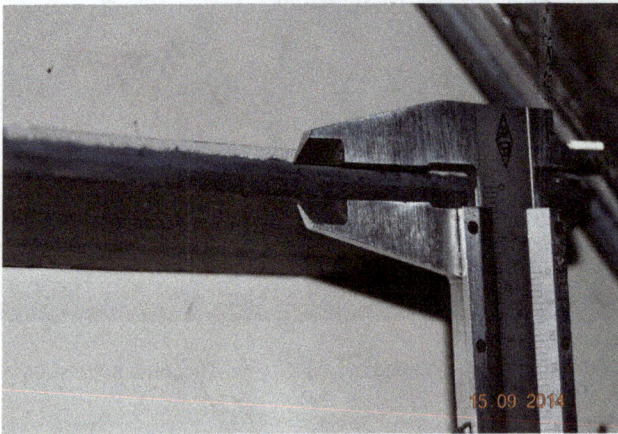

FIGURE 1.24   Measurement of a metal specimen's thickness with a vernier caliper after a break test.

Additionally, the weld beads must have a neat appearance:

- The three beads must share the same base width and height.
- Tie-ins must be well executed.
- Notches must be minimally visible.
- The throat thickness (for 5/16″ (8mm) plates) should be between 5/32″ and 1/4″ (4–6mm).

If the weld **meets all these requirements**, it is classified as **"acceptable"**. If **any criteria are not met**, it is considered **"not acceptable"**, and the **CSWIP WI** will provide **detailed feedback** to help the welder **improve for future tests** (Figure 1.26).

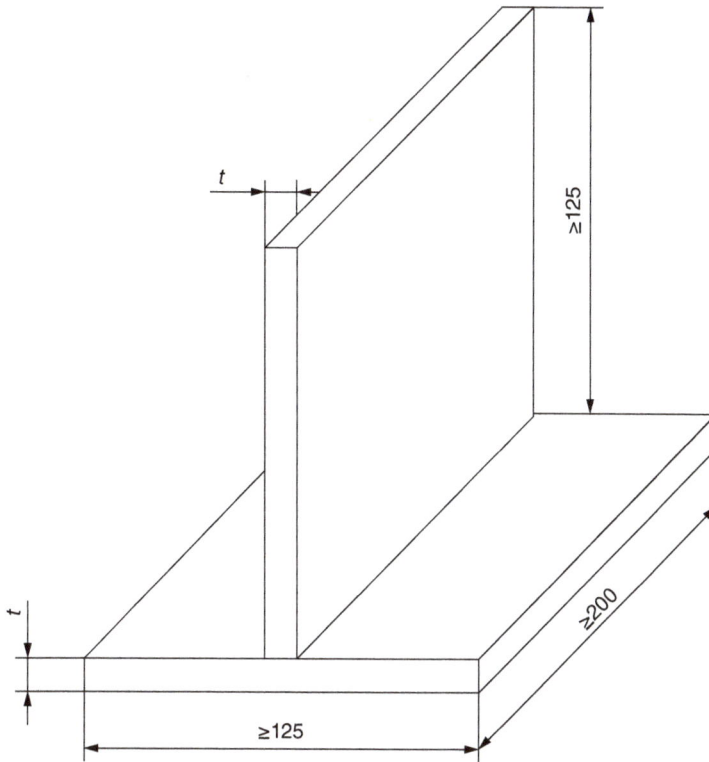

FIGURE 1.25 Dimensional diagram of a fillet joint specimen. The minimum plate dimensions and thickness are specified for a welding test.

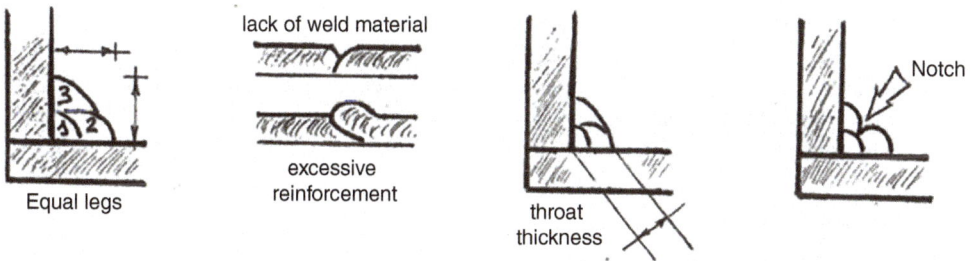

FIGURE 1.26 Welding diagrams of a fillet joint. They illustrate weld beads with equal leg lengths, defects in splices due to lack or excess of material, throat measurement, and notch formation.

SHIELDED METAL ARC WELDING

PRACTICE 8. FIRST WELD BEADS + OVERLAY IN VERTICAL-UP POSITION. PF (3G)

- **Base material:** Carbon steel plate 6″ x 6″ x 5/16″.
- **Electrodes to be used:**
  - o Ø 3/32″ x 14″ rutile electrode 6013. Direct polarity.
  - o Ø 3/32″ x 14″ basic electrode 7016. Reverse polarity.

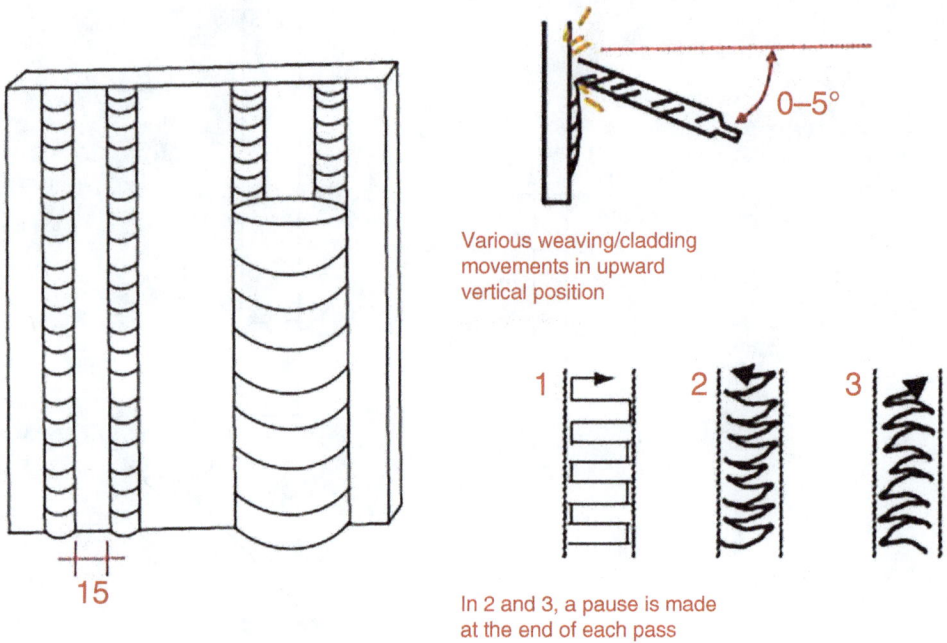

Various weaving/cladding
movements in upward
vertical position

In 2 and 3, a pause is made
at the end of each pass

FIGURE 1.27 Vertical-up welding with a shielded metal arc electrode, using weaving or buildup movements. On the right, diagrams show different travel patterns and electrode inclination.

- **Number of weld beads:** Six straight weld beads. Three overlay beads using a zigzag motion.
- **Welding current:**
  - o 60–70 amps for straight weld beads.
  - o 70–80 amps for overlay weld beads.

You are now attempting one of the most emblematic welding positions (Figure 1.27). Despite its difficulty, every skilled welder must master it. Additionally, this practice introduces the use of basic electrodes.

I. Straight weld beads

- Use the rutile electrode 6013.
- Mark the piece using a scribe and enhance visibility with a grinder and cutting disc or centre punch.
- Position the piece vertically.
- Select the recommended amperage.
- Hold the electrode perpendicular to the piece, slightly inclined at 5°.
- Maintain an arc length of 0.08–0.12″ (2–3mm) from the plate.
- Perform a zigzag or spiral motion while moving upwards. If you only move your elbow while keeping your arm supported against your body, you will control the motion better. Get used to using your body for support and use your other arm and legs to stabilise yourself with the surrounding elements.
- Ask the instructor if you have any questions.

II.   Overlay weld beads

- Fill the gap between the weld beads using an upward zigzag or U motion.
- Overlap each pass with the previous one.
- Let the slag cool before removing it with a chipping hammer to verify bead quality. Always wear PPE.

III.   Switching to basic electrode 7016

- Adjust the current according to the specified settings.
- Reverse polarity: Electrode to + and ground to -.
- This electrode requires a shorter arc and is more difficult to strike.
- Produces a superior finish with reduced sagging, as its thinner slag dissipates heat faster, allowing the metal to solidify sooner, preventing excessive fluidity.

ADDITIONAL TIPS

- ✓ If you need to stop while using a basic electrode, break the flux coating and let the core stick out slightly to restart more easily.
- ✓ Store the electrodes in an oven at 80–100°C immediately after opening the package to prevent moisture absorption.

💡 Moisture in the flux can cause porosity in the welds because the absorbed water evaporates, forming vapour pockets inside the weld bead. This occurs because the flux coating of basic electrodes is primarily composed of calcium carbonate, which naturally absorbs atmospheric moisture.

This document continues with structured welding exercises and practical applications. Let me know if you'd like additional refinements!

SHIELDED METAL ARC WELDING

PRACTICE 9. VERTICAL-UP FILLET WELD. PF (3F)

- **Base material:** Carbon steel plate. Two pieces 1.6″ x 6″ x 5/16″, tack-welded at 90°.
- **Electrodes to be used:**
  - o Ø 3/32″ x 14″ and Ø 1/8″ basic electrode 7016. Reverse polarity (electrode holder to + and ground clamp to -).
- **Number of weld beads:**
  - o Root pass (spiral or zigzag motion using Ø 3/32″ electrode).
  - o Weaving pass (lateral U motion using Ø 1/8″ electrode).
- **Welding current:**
  - o 70–80 amps for root and weave pass using Ø 3/32″ electrode.
  - o 80–95 amps for root or weave pass using Ø 1/8″ electrode.

INSTRUCTIONS

Position yourself on one side of the joint and make sure your head is above the end of the weld seam to avoid burns from spatter (Figure 1.28).

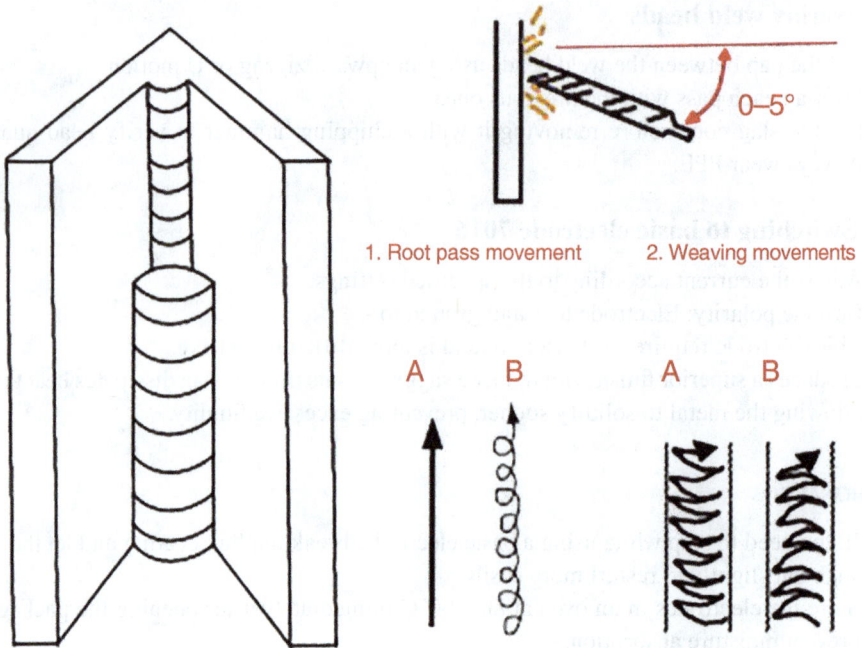

1. Root pass movement   2. Weaving movements

A   B   A   B

FIGURE 1.28   Welding in a vertical-up fillet joint with a shielded metal arc electrode. On the right, diagrams show the root pass movement and different weaving patterns for the filler layers.

1.   Root pass

- Use Ø 3/32″ basic electrode 7016.
- Perform an upward zigzag or spiral motion to "stitch" both sides of the fillet weld.
- Maintain a weld bead width equivalent to twice the electrode diameter.
- Use your wrist for movement and support your elbow against your torso for better control.
- Tilt the electrode approximately 5° above the horizontal plane (almost perpendicular to the joint).
- Keep the electrode tip 0.08–0.12″ (2–3mm) from the plate, ensuring it does not dip into the molten pool.
- Allow the slag to cool, remove spatter, and brush the weld before proceeding with the next pass.

2.   Weaving pass

- Use Ø 3/32″ or Ø 1/8″ basic electrode 7016.
- Perform a lateral U or zigzag motion, pausing for two seconds at each side of the root pass.
- Half of the electrode tip should remain outside the root pass, while the other half overlaps it.

WELDING MACHINE PARAMETER ADJUSTMENTS

- **Arc force:**
  o Slightly increase the arc force to prevent the electrode from sticking during striking and to improve penetration in the root pass.
  o A medium setting is ideal for controlling the molten pool without excessive reinforcement.
- **Inductance/dynamics:**
  o **Maintain a low inductance** to limit the size of the molten pool and avoid sagging due to gravity.
  o **Slightly increase the inductance** on the final pass to smooth the molten pool and enhance the transition between weld beads.

ADDITIONAL TIPS

✓ Ensure each weld bead is properly cleaned before proceeding with the next pass.
✓ Visually monitor bead progression and adjust your travel speed based on the molten pool behaviour.

With these settings, you will minimise sagging, achieve proper penetration, and produce well-profiled beads with smooth transitions in vertical-up welding.

SHIELDED METAL ARC WELDING

PRACTICE 10. OVERHEAD FILLET WELD. PD (4F)

- **Base material:** Carbon steel plate. Two pieces 1.6" x 6" x 5/16", tack-welded at 90°.
- **Electrodes to be used:** Ø 1/8" x 14" basic electrode 7016. Reverse polarity (electrode holder to +, ground to -).
- **Number of weld beads:** Three weld beads, executed with straight or slightly zigzag lateral movement.
- **Welding current:** 110–120 amps.

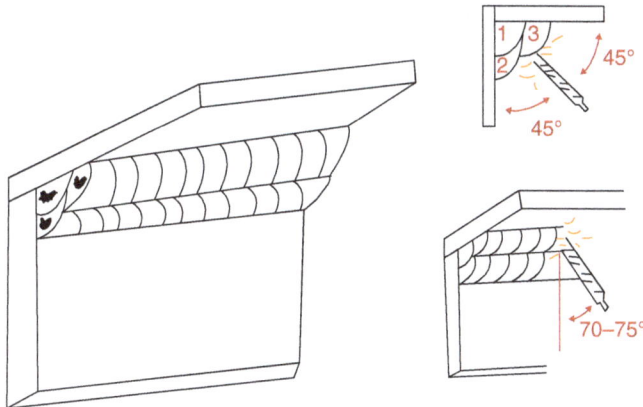

FIGURE 1.29 Overhead welding in a fillet joint with a shielded metal arc electrode. On the right, diagrams show the bead deposition sequence, electrode inclination angle, and travel direction.

INSTRUCTIONS

When welders take a qualification test in this position (PD) or vertical-up (PF), it is usually because they know that passing this test will also certify them for horizontal fillet welding (Figure 1.29). Additionally, if they use a basic electrode, they will also be qualified to weld with rutile electrodes. These PF and PD qualification tests are essential for welding metal structures in construction.

1. Preparation

- Position yourself on one side of the joint, ensuring the weld is at eye level to avoid burns from spatter.
- Make sure the edges of both plates align perfectly before tacking.
- If you notice any gaps allowing light through, grind the edges with a grinding disc until they are perfectly smooth.

2. Root pass

- Use Ø 1/8″ basic electrode 7016.
- Hold the electrode at a 45° angle to the joint, with a 70°–80° inclination in the direction of travel.
- Perform a straight motion or a slight lateral zigzag.
- Maintain an arc length of 0.08–0.12″ (2–3mm) from the plate.

3. Weaving passes

- Execute the second weld bead aiming at the root pass to completely cover it.
  o Avoid leaving the bead concentrated only on the lower plate.
- The third weld bead should aim at the upper section of the root pass.
- Maintain the same angles and arc length throughout the process.

WELDING MACHINE PARAMETER ADJUSTMENTS

- **Arc force:**
  o Slightly increase the arc force to prevent the electrode from sticking during arc striking and to enhance penetration in the root pass.
  o A medium setting is ideal to control the molten pool without excessive reinforcement.
- **Inductance/dynamics:**
  o Maintain a low inductance to limit the size of the molten pool and prevent sagging due to gravity.
  o Slightly increase the inductance in the final pass to smooth the molten pool and improve the transition between weld beads.

⚒ **Remember:** These settings are only suggestions. Test, adjust, and personalise them to match your own welding style. You are part of the craftsmanship that shapes each weld bead.

ADDITIONAL TIPS

- ✓ Maintain a steady travel speed, not faster than in previous exercises. The recommended bead width should be 1.5 to 2 times the electrode diameter.
- ✓ Wear PPE and follow all safety measures.
- ✓ Weld with confidence, trusting in your ability to achieve the best possible result.

Did you succeed? Give me your best smile!

SHIELDED METAL ARC WELDING

PRACTICE 11. LAP JOINT WELDING

- **Base material:** Carbon steel plate. Two pieces 1.6" x 4" x 1/8".
- **Electrodes to be used:**
  - o Ø 3/32" x 14" rutile electrode 6013. Direct polarity.
  - o Ø 3/32" x 14" basic electrode 7016. Reverse polarity (electrode holder to +, ground to -).
- **Number of weld beads:**
  - o One weld bead, executed with a straight movement.
- **Welding current:** 65–75 amps (Figure 1.30).

1. Preparation

- Tack-weld the plates together in a lap joint with strong tacks.
- Ensure the plates are properly aligned and firmly secured.

2. Welding execution

- Use Ø 3/32" rutile electrode 6013 on one side and basic electrode 7016 on the opposite side.
- Position the electrode laterally to the joint with a 45° inclination.
- Maintain a 70°–80° inclination in the direction of travel.
- Keep an arc length of 0.08–0.12" (2–3mm) from the joint.
- Adjust the travel speed to ensure the molten pool extends at least 0.04" (1mm) beyond the edge of the upper plate.
- The weld bead must maintain a consistent width from start to finish and fully fuse the edge of the upper plate.
- If, after welding, the upper plate's edge is visible in any area, it indicates that the travel speed was too fast.

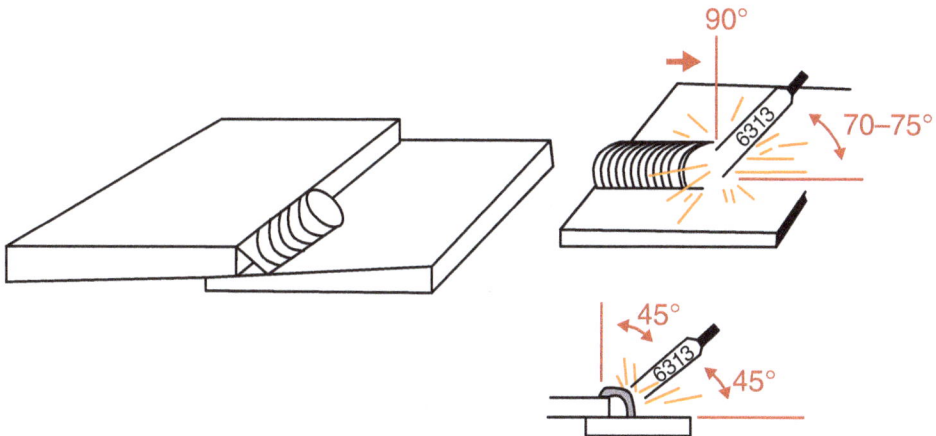

FIGURE 1.30   Welding in a lap joint with a shielded metal arc electrode. On the right, diagrams show the electrode inclination angles and travel direction in horizontal and angled positions.

ADDITIONAL TIPS

✓ Maintain a steady travel speed and ensure that the weld bead fully covers the "step" formed by the overlapped upper plate on the lower plate.

SHIELDED METAL ARC WELDING

PRACTICE 12. V-GROOVE PLATES IN HORIZONTAL POSITION. PA (1G)

- **Base material:** Carbon steel plate. Two pieces 6" x 1.6" x 5/16".
- **Electrodes to be used:** Ø 3/32" x 14" basic electrode 7016 for root pass, filler passes, and capping pass. Reverse polarity.
- **Number of weld beads:**
  o 1st (root pass): Movement of choice: zigzag, back-and-forth, or circular.
  o 2nd and 3rd (filler passes): Two weld beads with straight motion.
  o 4th (capping pass): Lateral U or Z motion.
- **Welding current:**
  o 50–60 amps for root pass.
  o 80–90 amps for filler and capping passes.

You are now tackling the first welder qualification position with V-groove edge preparation. This method is used for thicknesses between 5/32" and 1/2" (4–12mm) when the welder has access to only one side of the piece, as in the welding of small pipes. The success of this practice depends 50% on your skill and 50% on proper preparation of the piece (Figure 1.31).

1. Preparation

- Bevel the edges of the plates at 35°, considering the plate thickness.
  o For electrode welding, always bevel full-thickness butt joints from 5/32" (4mm) upwards to ensure full penetration.
- Once bevelled, grind the edge of the groove to 2–3mm (known as the "land") to prevent excessive burn-through.

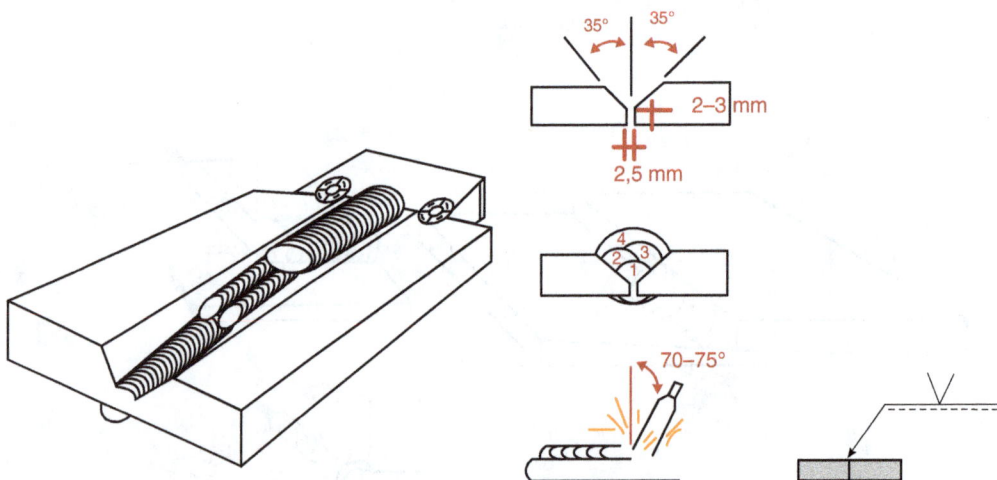

FIGURE 1.31 Welding in a V-groove butt joint using a shielded metal arc electrode. On the right, diagrams show the joint preparation, root bead penetration, and electrode inclination angle.

- Position both plates with a 3/32" (2.5mm) root gap, ensuring perfect alignment.
  - o Use two backing strips (small steel pieces, one at each end).
  - o Tack-weld them to the plates once aligned to maintain their position throughout welding.
  - o Backing strips also allow arc striking and weld termination outside the actual joint, reducing defects at the start and end.
- Elevate the joint slightly using the backing strips to prevent direct contact with the worktable.

2. Root pass

- Use basic electrode 7016 with reverse polarity (electrode to positive terminal).
- Strike the arc on the backing strip, then slowly move along one bevel edge until reaching the start of the weld joint.
- Hold the electrode at a 75°–70° inclination and maintain a 0.08–0.12" (2–3mm) arc length.
- Control the size of the "keyhole" using a chosen movement: zigzag, back-and-forth, or circular.
  - o This ensures fusion of the root gap while preventing excessive keyhole enlargement.
- Arc force: Set high to ensure penetration and prevent electrode sticking.
- Inductance: Set to minimum for controlled fusion and a stable molten pool.
- Tie-in for root pass (see Figure 1.32):
  - o Grind the last 3/4" (20mm) of the weld bead with a grinder to form a ramp.
  - o Restart the arc on the ramp and remelt its end to achieve a seamless tie-in.

3. Filler passes

- Set welding current to 80–90 amps.
- Apply two weld beads side by side, filling the gap above the root pass.
- Practise tie-ins as done in Practice No. 3.

4. Capping pass

- Execute the capping pass with a U or zigzag motion along the groove lines.
- Keep the electrode within the bevel lines, pausing momentarily at the sides to deposit enough filler material.
- Arc force: Maintain medium settings to reduce spatter and ensure proper fusion.
- Inductance: Increase slightly to smooth the molten pool and achieve a clean, flat finish.
- Overlap each U or zigzag movement with the previous one to ensure uniform bead profile.
- The final weld reinforcement should not exceed 2–3mm in height.

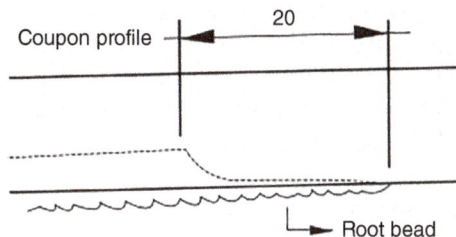

FIGURE 1.32   Weld coupon profile illustrating the grinding technique used at the root bead to guarantee fusion.

Try performing the root pass with direct polarity (electrode to negative terminal) and observe the following effects:

- ✓ **Reduced penetration:** Direct polarity produces shallower penetration, making it harder to reach the root if the gap is large.
- ✓ **Less spatter:** A more controlled molten pool, ideal for cleaner welds in visible areas.
- ✓ **More stable arc:** The electrode heats up less, allowing greater control over manipulation, but requiring greater precision to avoid defects.

ADDITIONAL TIPS

- ✓ Maintain a steady travel speed and adapt techniques as needed to prevent defects.
- ✓ Wear PPE and follow all safety measures.

Every weld bead you create is another step towards mastering welding. Don't fear mistakes—they are lessons in disguise. Recognise your progress with humility and refine your skills with patience.

 ⚒ What seems difficult today will be just another milestone tomorrow. Keep going with determination and confidence!

EVERY WELD BEAD YOU LAY IS A STEP TOWARDS MASTERING WELDING.

Do not fear making mistakes: every imperfection is a hidden lesson. Recognise your progress with humility and work patiently to refine your craft. What seems difficult today will be just another milestone overcome tomorrow. **Keep going with determination and confidence!**

INSPECTION PROTOCOL FOR **CSWIP** IN FULL
PENETRATION **V**-GROOVE BUTT WELDS

When performing a full penetration V-groove butt weld, the role of the CSWIP is crucial in ensuring that the weld meets quality and safety standards. The following outlines the typical steps followed by a CSWIP inspector during an inspection:

1. Visual inspection

- ✓ The CSWIP inspector examines the weld to identify visible surface defects, such as cracks, porosity, undercut, lack of fusion, or excessive reinforcement.
- ✓ Ensures that the weld bead profile is uniform and free from discontinuities.
- ✓ Visual inspection is the first step, which helps identify obvious issues that require correction before proceeding with further testing.

2. Dimensional measurements

- ✓ Measures the dimensions of the weld bead, including throat thickness, reinforcement height, and joint alignment.
- ✓ Verifies that these dimensions match the specifications outlined in the welding procedure.

3. Non-destructive testing (NDT)

✓ **Radiographic testing (RT):**
  • Uses X-rays to inspect the internal structure of the weld, detecting porosity, slag inclusions, or cracks that are not visible externally.
✓ **Ultrasonic testing (UT):**
  • Uses high-frequency sound waves to detect internal discontinuities, such as lack of fusion or slag inclusions.
✓ **Dye penetrant testing (PT):**
  • A coloured liquid is applied to the weld surface, which reveals fine cracks or porosity when wiped off and exposed to a developer.
✓ **Magnetic particle testing (MT)** (for **ferromagnetic materials**):
  • Uses fine magnetic particles to detect cracks on or just beneath the weld surface.

4. Destructive testing

✓ **Tensile test:**
  • Assesses the weld's resistance to tensile stress until failure occurs.
  • Determines the strength of both the weld metal and the parent material.
✓ **Bend test:**
  • Evaluates the ductility and strength of the welded joint.
  • The specimen is bent to check for cracks or separations.
✓ **Charpy impact test:**
  • Measures the toughness of the weld metal at low temperatures.
  • Determines the energy absorbed by the specimen before fracture.
✓ **Macroscopic examination (macroetch test):**
  • Involves cutting, polishing, and chemically etching a cross-section of the weld.
  • Reveals penetration depth and overall weld quality.

5. Evaluation of results

✓ All tests must meet the acceptance criteria defined in relevant standards (such as ISO 5817 or AWS D1.1, depending on project requirements).
✓ If a test is deemed "non-acceptable", the weld does not comply with the required standards.
  • Depending on the severity of the defect and project specifications, various corrective actions may be taken:

✑ Repair: In some cases, the defective weld may be repaired and re-tested.
  ✑ Rejection: If the defects cannot be repaired, or the weld does not meet critical specifications, the joint may be rejected and must be re-welded.

6. Documentation

✓ The CSWIP inspector records all observations and test results, providing a detailed inspection report.
✓ The report includes measurements, test outcomes, and recommendations for corrective actions if required.

If you wish to practise this test on a qualification-sized coupon, the minimum dimensions (in millimetres, Figure 1.33) according to **UNE EN ISO 9606–1** are:

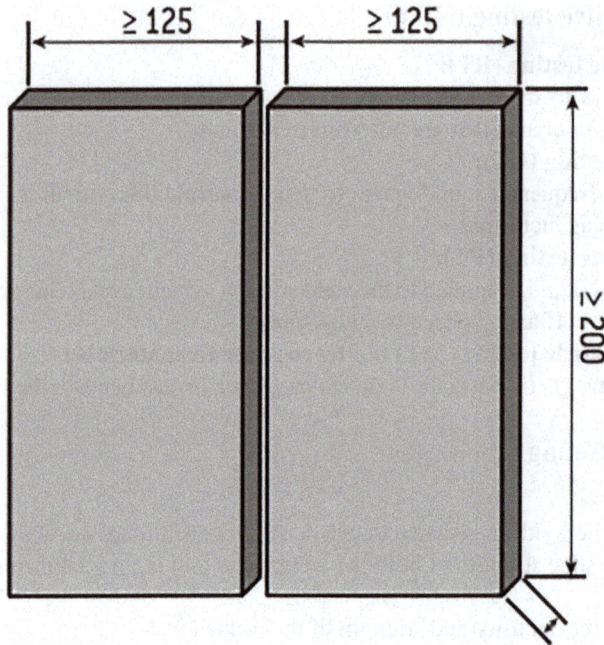

FIGURE 1.33   Dimensional diagram of a butt joint specimen. The minimum plate dimensions are indicated.

SHIELDED METAL ARC WELDING

PRACTICE 13. V-GROOVE PLATES IN HORIZONTAL-VERTICAL POSITION. (PC-2G)

- **Base material:** Carbon steel plate. Two pieces 6″ x 1.6″ x 5/16″.
- **Electrodes to be used:** Ø 3/32″ x 14″ basic electrode 7016 for root pass, filler passes, and capping pass (reverse polarity).
- **Number of weld beads:** Six weld beads, all executed with a straight motion.
- **Welding current:**
  - o   50–60 amps for root pass.
  - o   80–90 amps for filler and capping passes.

INSTRUCTIONS

After working on the previous practice, this time we will focus only on the aspects that differ in the PC position (horizontal-vertical fillet weld, Figure 1.34).

1.   Plate preparation

- √ The bevel angles may vary:
  - • 40°–45° for the upper edge.
  - • 25° for the lower edge.
  - • This configuration helps support the weld pool, although it is not essential.
- √ Ensure the edges of the plates are properly bevelled and aligned before starting.

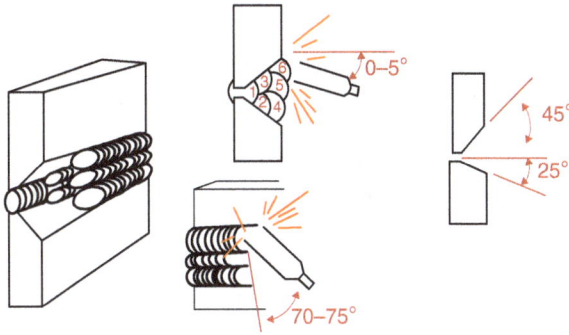

FIGURE 1.34 Ledge position butt welding with a V-groove using a shielded metal arc electrode. On the right, diagrams show the joint preparation, bead deposition sequence, and electrode inclination.

2. Root pass

✓ Direct the electrode slightly towards the lower bevel.
  • Heat rises naturally, causing the upper bevel to melt faster if the electrode is aimed equally at both edges.
✓ Control the "keyhole" using a straight, lateral, or circular motion, as explained in the previous practice.
✓ Maintain the correct electrode inclination and arc length to ensure proper penetration.
✓ Arc force:
  • Set to medium-high to ensure good penetration in the root pass.
  • Prevents electrode sticking.
✓ Inductance:
  • Keep low to stabilise the weld pool and minimise the risk of excessive molten metal flow.

3. Filler passes

✓ Apply the filler passes (2nd and 3rd beads) with a straight, lateral, or circular motion.
✓ Always start with the lower bead and follow the sequence shown in the figure to ensure uniform filling.

4. Capping passes

✓ For the capping passes (4th, 5th, and 6th beads), follow the same method as in Practice No. 5.
✓ Lay the three passes in sequence, starting with the lowest, then the middle, and finally the top.
✓ Practise tie-ins to ensure smooth transitions between welds without defects.
✓ Arc force:
  • Keep at medium to reduce spatter and achieve clean fusion.
✓ Inductance:
  • Increase slightly to smooth the molten pool and improve the final appearance.

ADDITIONAL TIPS

✓ Stay focused and follow the recommendations from the previous practice.
✓ Ensure each weld bead is well aligned and maintains the correct thickness.
✓ Wear PPE and follow all safety guidelines.
✓ Trust in your skill and the knowledge you've acquired so far.

Every step is a lesson. Learn to adjust, control, and refine your technique with every weld bead you lay. Acknowledge your progress and keep moving forward with humility and perseverance. What seems like a challenge today will become a mastered skill tomorrow. Keep practising with determination and confidence!

SHIELDED METAL ARC WELDING

PRACTICE 14. V-BEVELLED PLATES – VERTICAL UP POSITION. PF(3G).

- **Base material:** Carbon steel flat bar. Two pieces 6″ × 1.6″ × 5/16″.
- **Electrodes to be used:**
  o Ø 3/32″ × 14″ basic electrode 7016 for root pass (reverse polarity).
  o Ø 3/32″ or 1/8″ × 14″ basic electrode 7016 for filler and capping passes (reverse polarity).
- **Number of weld beads:**
  o Root pass: Straight motion with basic electrode, slightly pushing the molten pool with the tip to ensure full penetration.
  o Filler pass: Lateral weaving with 3/32″ basic electrode. In some cases, one root pass and a cap may be sufficient.
  o Capping pass: "U"-shaped motion with 3/32″ or 1/8″ basic electrode.
- **Welding current:**
  o 50–60 amps for root pass.

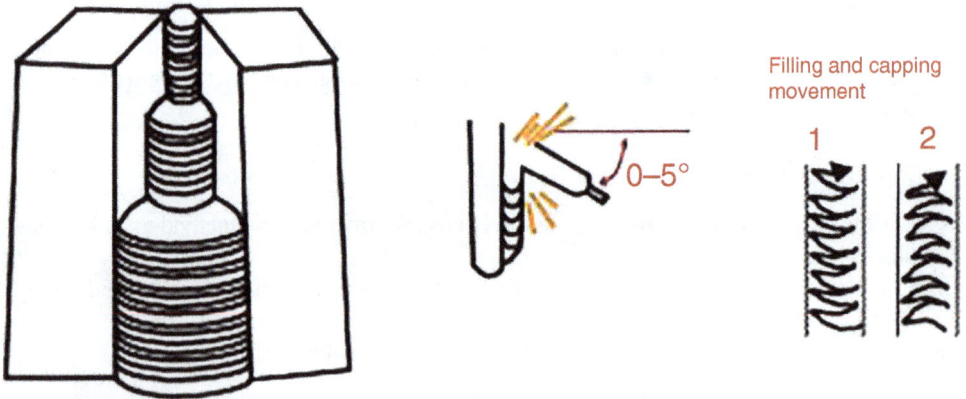

Filling and capping movement

0–5°

1   2

FIGURE 1.35   V-groove welding in the vertical-up position with a shielded metal arc electrode. On the right, diagrams show the electrode inclination angle and the weaving and filling movements in welding.

- o   65–80 amps for filler pass.
- o   70–80 amps with 3/32″ electrode; 95–100 amps with 1/8″ electrode for capping pass.

INSTRUCTIONS

This practice focuses on specific aspects of vertical-up welding, following the same procedure as Practice 11 for all other details (Figure 1.35).

1.   Root pass

- ✓ Hold the electrode at a 5°–10° inclination from horizontal.
- ✓ As you progress upwards, push the electrode into the joint until the arc light is visible on the back side.
  - • At that moment, pull back and repeat the technique along the entire joint.
  - • ✓ In some cases, simply moving upwards may be enough to close the root without additional movement.
  - • ✓ If a large "keyhole" forms, use a slight lateral motion to control it.
- ✓ Arc force:
  - • Set medium-high to ensure full penetration and prevent electrode sticking.
- ✓ Inductance:
  - • Set to minimum to stabilise the molten pool without causing excessive sagging.

2.   Filler pass

- ✓ Apply a lateral motion, pausing slightly at the edges of the root pass.
- ✓ Maintain the same electrode inclination as in the root pass.
- ✓ Arc force:
  - • Medium setting to prevent excessive molten pool flow.
- ✓ Inductance:
  - • Slightly increased to improve molten pool fluidity.

3.   Capping pass

- ✓ Use a U-shaped lateral motion, ensuring you do not exceed the groove lines.
- ✓ Control the amount of deposited material; the weld bead should not extend more than 1–2mm beyond the bevel edge.
- ✓ Arc force:
  - • Set to medium-high for proper fusion.
- ✓ Inductance:
  - • Increased to enhance final bead appearance.

ADDITIONAL TIPS

- ✓ Practise tie-ins to ensure a uniform and strong joint.
- ✓ Maintain the same quality and precision as in Practice 11, so both welds appear virtually identical.
- ✓ Always wear PPE and follow all safety protocols.
- ✎ Keep going! Practise until you achieve a clean, well-finished weld bead.

SHIELDED METAL ARC WELDING

PRACTICE 15. V-GROOVE PLATES IN 45° INCLINED VERTICAL-UP POSITION (PF-3G)

- **Base material:** Carbon steel plate. Two pieces 6″ × 1.6″ × 5/16″.
- **Electrodes to be used:**
  - o Ø 3/32″ × 14″ basic electrode 7016 for root pass (reverse polarity).
  - o Ø 1/8″ × 14″ basic electrode 7016 for filler and capping passes (reverse polarity).
- **Number of weld beads:**
  - o Root pass: Straight motion, slightly pushing the molten pool with the electrode tip to ensure full penetration.
  - o Filler pass: Lateral movement (with Ø 3/32″ basic electrode). In some cases, a single root pass followed by capping may be sufficient.
  - o Capping pass: U-shaped motion (with Ø 1/8″ basic electrode).
- **Welding current:**
  - o 50–60 amps for root pass.
  - o 65–80 amps for filler pass.
  - o 70–80 amps for Ø 3/32″ and 95–100 amps for Ø 1/8″ in capping pass.

1. Positioning and execution

✓ The test coupon may be inclined at 45° to the left or right (Figure 1.36).
✓ If the coupon is inclined to the left and you are right-handed:
  - Position your head at the upper left corner to get a clear view.
  - Execute the root pass in the same way as in the previous practice.
✓ If the coupon is inclined to the right:
  - Position your head near the upper right corner and use your left hand to guide the electrode.
  - This is an excellent opportunity to start practising with both hands, which is crucial for advancing your welding skills.
  - If you are left-handed, do the opposite.

2. Capping pass

✓ Use either your left or right hand depending on the inclination of the test coupon.

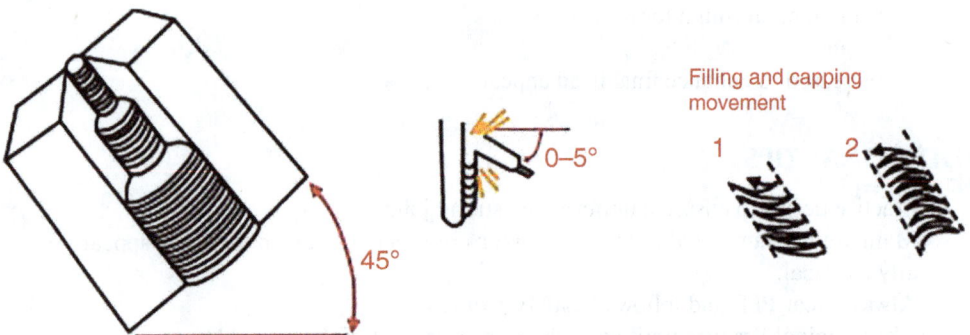

FIGURE 1.36 V-groove welding in the vertical-up position (45°) with a shielded metal arc electrode. On the right, diagrams show the weaving and filling movements in welding.

✓ You can opt for a lateral or U-shaped motion, as explained in the previous practice.
✓ Ensure that the capping pass remains parallel to the ground to prevent undercutting on the bevel edges.

Mastering this technique will help improve your adaptability and precision in different welding positions. Keep practising with patience and confidence!

SHIELDED METAL ARC WELDING

PRACTICE 16. V-GROOVE PLATES IN OVERHEAD POSITION. (PE-4G)

- **Base material:** Carbon steel plate. Two pieces 6″ × 1.6″ × 5/16″.
- **Electrodes to be used:**
  - Ø 3/32″ × 14″ basic electrode 7016 for root pass and filler passes (reverse polarity).
  - Ø 1/8″ × 14″ basic electrode 7016 for capping pass (reverse polarity).
- **Number of weld beads:**
  - Root pass: Straight motion (with Ø 3/32″ basic electrode), ensuring deep penetration into the joint.
  - Filler passes: Straight motion (with Ø 3/32″ basic electrode).
  - Capping pass: U-shaped lateral motion (with Ø 3/32″ or Ø 1/8″ basic electrode).
- **Welding current:**
  - 50–60 amps for root pass.
  - 65–80 amps for filler passes.
  - 70–80 amps for Ø 3/32″ and 95–100 amps for Ø 1/8″ in capping pass (Figure 1.37).

SUCCESS IN THIS CHALLENGING POSITION DEPENDS MORE THAN EVER ON SEVERAL KEY FACTORS

✓ Root face thickness ("land")
✓ Root gap distance
✓ Properly adjusted welding current

Crucial adjustment:
- If the current is too high → The root face will burn through, leaving a concave root bead.
- If the current is too low → The root will not fuse properly, leaving unfused sections in the joint.

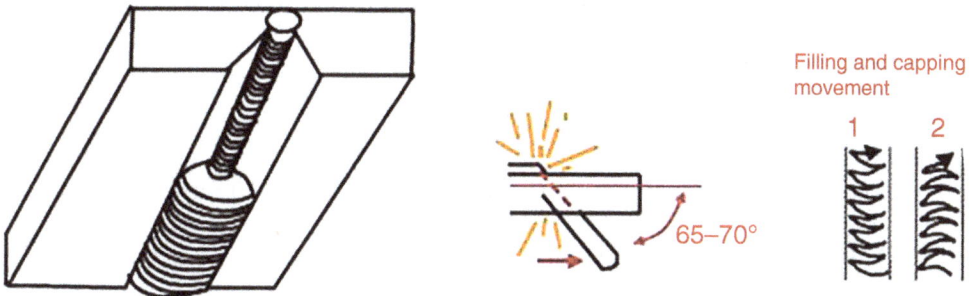

FIGURE 1.37 V-groove welding in the overhead position with a shielded metal arc electrode. On the right, diagrams show the electrode inclination angle and the weaving and filling movements in welding.

1. Root pass

  ✓ Hold the electrode at a 65°–70° inclination and push it into the joint.
  ✓ Perform multiple test runs, adjusting the current slightly up or down (by 2–3 amps).
  ✓ The root bead on the upper side must remain flat but never concave.
  ✓ Arc force:
    • Medium-high to ensure penetration without excessive sagging.
  ✓ Inductance:
    • Low, to control the molten pool and prevent excessive shrinkage.

2. Filler passes

  ✓ Use a straight motion with basic electrodes.
  ✓ Adjust the current according to electrode diameter.
  ✓ Arc force:
    • Medium, to maintain control without excessive burning.
  ✓ Inductance:
    • Medium, to maintain fluidity and facilitate filling.

3. Capping pass

  ✓ Use a U-shaped motion, pausing at the sides and moving quickly across the centre.
  ✓ Keep the electrode within the bevel lines to ensure full fusion.
  ✓ Arc force:
    • Medium-high, to ensure complete fusion and prevent porosity.
  ✓ Inductance:
    • Medium, to maintain fluidity and facilitate filling.
  ✓ Experiment with different settings; the key is not to settle for the first result.

Mastering overhead welding requires patience, precision, and continuous practice. Keep refining your technique, and every weld will bring you closer to mastery!

SHIELDED METAL ARC WELDING

PRACTICE 17. X-GROOVE PLATE IN HORIZONTAL-VERTICAL POSITION (PC-2G)

  • **Base material:** Carbon steel plate. Two pieces 6″ × 1.6″ × 1–3/16″.
  • **Electrodes to be used:**
    o Ø 1/8″ × 14″ basic electrode 7016 (reverse polarity).
    o Ø 5/32″ × 14″ basic electrode 7016 (reverse polarity).
  • **Number of weld beads:** Six to ten passes per side:
    o Root pass: Straight motion.
    o Filler and capping passes: Straight motion.
  • **Welding current:**
    o Root pass: 85–90 amps for the first root pass (increase by 5–10 amps for the second root pass).
    o Filler and capping passes: 95–105 amps for Ø 1/8″ electrode, 110–130 amps for Ø 5/32″ electrode.

This exercise focuses on performing a full penetration weld on thick-section plates where the welder has access to both sides of the workpiece (Figure 1.38). This type of joint requires specific edge preparation using an "X" bevel.

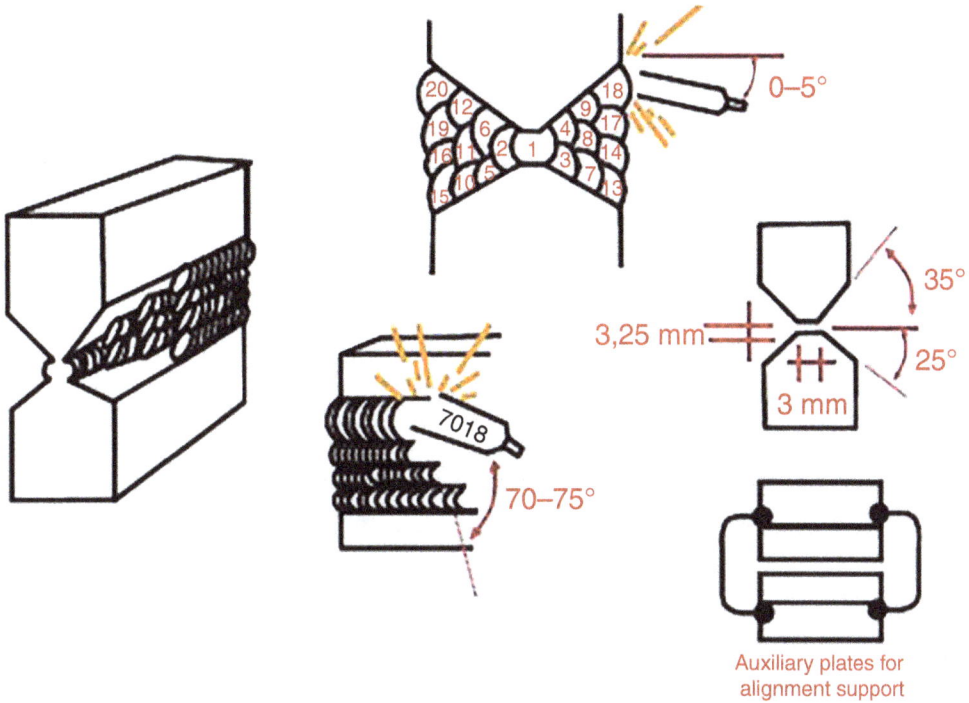

FIGURE 1.38 X-groove welding in the vertical-up position with a shielded metal arc electrode. On the right, diagrams show the joint preparation, bead deposition sequence, electrode inclination, and the use of bridges to maintain alignment.

1. Workpiece preparation

✓ Bevel both edges symmetrically:
- 25° on the bottom.
- 35° on the top.

✓ Leave a 1/8″ (3mm) root face and a 1/8″ (3.25mm) root gap.

✓ Insert a stripped Ø 1/8″ (3.25mm) electrode bent into a V shape into the joint to maintain separation during tacking.

2. Root pass (Side A)

✓ Position the test coupon in the horizontal-vertical position and select reverse polarity.

✓ Strike the arc on the tack weld and, once stabilised, move the electrode into the root.

✓ Use the same techniques as in previous exercises to control fusion, keeping the electrode inclined at 70–75° in the travel direction.

3. Root pass (Side B)

✓ Flip the test coupon over and repeat the process on "Side B".

✓ Before welding, grind down the root penetration from the first bead using a grinder to remove any excessive material.

✓ Increase the welding current by 5–10 amps to ensure strong fusion between the second root bead and the first.

4. Filler and capping passes

✓ Return to "Side A" and set the appropriate polarity and current for the electrode used.
✓ Deposit two filler passes: One below the first bead and another above it.
✓ Repeat the process on "Side B".
✓ Now, deposit three more passes on "Side A" and three more on "Side B", alternating sides to balance stress and prevent distortion.
✓ If the bevel lines are still visible, continue welding until the joint is fully closed, adding additional passes as necessary.

FINAL RECOMMENDATIONS

✓ Welding sequence: Follow a specific welding order to balance expansion and contraction, preventing plate deformation. Do not remove tack welds until the weld is complete.
✓ Temperature control: As welding progresses, adjust the current if you notice excessive heat build-up in the test coupon.

This exercise requires patience and careful planning to achieve a perfect result. Keep practising with confidence and determination!

SHIELDED METAL ARC WELDING

PRACTICE 18. X-GROOVE PLATE IN VERTICAL-UP POSITION (PF-3G)

• **Base material:** Carbon steel plate. Two pieces 6″ × 1.6″ × 1–3/16″ butt-jointed.
• **Electrodes to be used:**
   o Ø 1/8″ × 14″ basic electrode 7016 (reverse polarity).
   o Ø 5/32″ × 14″ basic electrode 7016 (reverse polarity).
• **Number of weld beads:** Five to eight passes per side:
   o Root pass: Straight motion.
   o Filler and capping passes: Upward zigzag or U motion.
• **Welding current:**
   o Root pass: 85–90 amps for the first root pass (increase by 5–10 amps for the second root pass).
   o Filler and capping passes: 95–105 amps for Ø 1/8″ electrode, 110–130 amps for Ø 5/32″ electrode.

This exercise focuses on performing a vertical-up weld using an "X" bevel preparation (Figure 1.39). Maintaining an alternating sequence between Side A and Side B is crucial to achieving uniform heat distribution and preventing distortion.

1. Workpiece preparation

✓ Bevel the joint at 30° on both sides.
✓ Maintain a 1/8″ (3mm) root face and a 1/8″ (3.25mm) root gap.
✓ Ensure both pieces are perfectly aligned and securely clamped before welding.

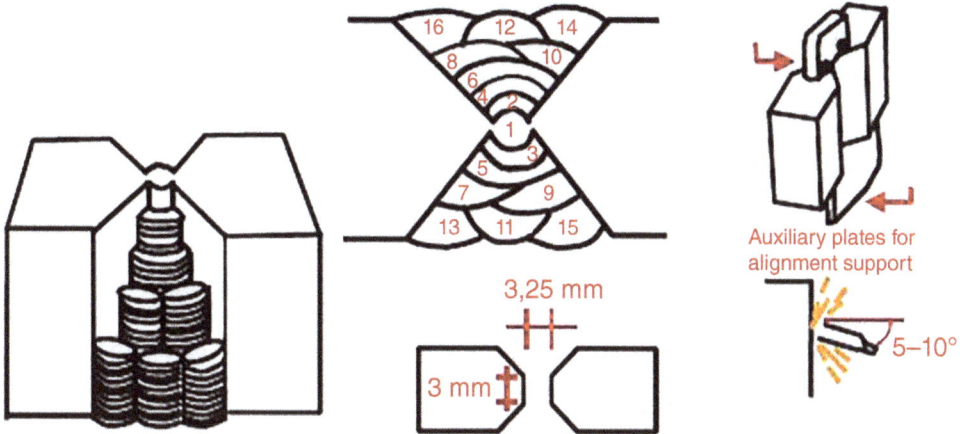

FIGURE 1.39 X-groove welding in the vertical-up position with a shielded metal arc electrode. On the right, diagrams show the joint preparation, bead deposition sequence, electrode inclination, and the use of bridges to maintain alignment.

2. Root pass (Side A and B)

✓ Apply the same techniques used in Exercise No. 14 with the basic 7016 electrode until the root pass on Side A is completed.

✓ Repeat the process on Side B. Increase the current by 5–10 amps when welding the second root pass to ensure full fusion with the first.

3. Filler passes

✓ Passes 2 and 3 (Side A and B): Use a U motion to fill the weld joint.

✓ From Pass 4 onwards (Side A and B): The gap to be filled will be wider, requiring two overlapping U motion filler passes, one on each side of the joint.

4. Capping passes

✓ Final passes: If necessary, deposit three capping passes—one in the centre and two additional passes, one on either side.

5. Alternating sides

✓ Alternating weld passes between Side A and Side B is essential. This distributes heat evenly, minimising the risk of warping.

✓ If distortion is still an issue, consider switching sides after each pass.

FINAL RECOMMENDATIONS

✓ Refer back to Exercise No. 14 for detailed guidance on vertical-up welding techniques.

✓ Maintain consistency in movement and side alternation to achieve a professional-quality finish.

By following these guidelines, you will develop the skills required to achieve strong and well-finished welds in a vertical-up position. Keep practising and refining your technique!

SHIELDED METAL ARC WELDING

PRACTICE 19. WELDING A 2-INCH PIPE (ROTATING WHILE WELDING). PA(1G)

- **Base material:** Two 2″ round carbon steel pipes, 1/8″ wall thickness, 1–9/16″ long (commonly used in gas and fire protection systems, Figure 1.40).
- **Electrodes to be used:** Ø 3/32″ × 14″ rutile electrode 6013 (reverse polarity!).
- **Number of weld beads:** One weld bead with a lateral motion, keeping the electrode always positioned at the highest point of the pipe while rotating the pipe with the other hand.
- **Welding current:** 50 amps.

INSTRUCTIONS

1. Pipe preparation

✓ Remove burrs from the cut edges using a file to ensure a proper fit.
✓ Tack-weld the pipes, ensuring perfect alignment. It is crucial that there is no gap between the pipes and that no light passes through the joint.

2. Positioning

✓ Set the welding machine to the recommended current.
✓ Place the pipes on the workbench so that the pipe opening faces you.
✓ Attach the ground clamp to the highest point of the pipe.
✓ Hold the ground clamp with your left hand and the electrode holder with your right hand, pointing at the weld seam at the highest point of the pipe (for right-handed welders).

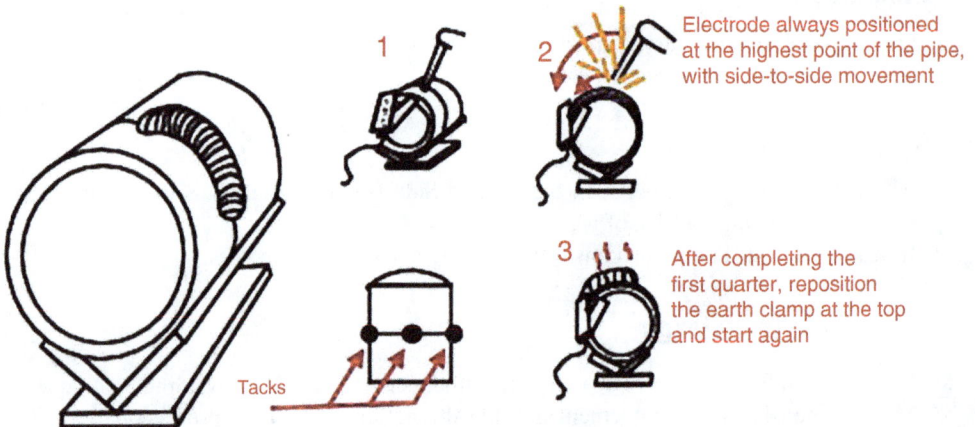

1

2   Electrode always positioned at the highest point of the pipe, with side-to-side movement

3   After completing the first quarter, reposition the earth clamp at the top and start again

Tacks

FIGURE 1.40 Circumferential welding on a pipe using a shielded metal arc electrode. On the right, diagrams show the quartered welding sequence, the lateral movement of the electrode, and the need to reposition the ground clamp after completing each section.

3. Welding process

✓ Start welding at the highest point of the pipe, moving the electrode laterally across the joint.

✓ While welding, rotate the pipe with your left hand counterclockwise, ensuring that the weld is always performed at the highest point.

✓ Weld a quarter of the pipe, then stop, reposition the pipe, and repeat the process until the weld is completed. Be careful! Excessive current can burn through the pipe wall. Pay close attention to the molten pool, as it provides a warning before burn-through occurs. Pause as many times as needed to control penetration.

4. Tips for tie-ins

✓ Follow the tie-in techniques described in Exercise No. 2.

✓ Maintain a steady pace, allowing the molten metal to penetrate the joint properly.

✓ The weld bead should be uniform, approximately 3/16″ (5mm) wide and no more than 1/8″ (3mm) high.

FINAL CONSIDERATIONS

This might seem like a challenging task for your first pipe welding experience, but if you have successfully completed the previous exercises, **you will master this technique in no time**. **Take your time, stay focused, and prioritise consistency and quality in each section of the pipe**.

WELDING STAINLESS STEEL AISI 304-L WITH SHIELDED METAL ARC WELDING (SMAW) USING AWS 308-L ELECTRODES

AISI 304-L stainless steel is one of the most widely used materials due to its properties, making it suitable for a wide range of applications. This type of stainless steel is an iron-based alloy containing a significant amount of chromium (between 12% and 30%), which provides excellent resistance to oxidation. Additionally, elements such as molybdenum, titanium, and nickel improve its tensile strength and weldability.

The welding technique for stainless steel is quite similar to that used for carbon steel; however, there are some key additional considerations due to the specific characteristics of stainless steel and the type of electrode used.

KEY CONSIDERATIONS FOR WELDING STAINLESS STEEL

1. Cleaning and preparation

✓ **Extreme cleanliness:** Stainless steel is highly susceptible to contamination, which can compromise its corrosion resistance. It is crucial to use dedicated tools exclusively for stainless steel, such as stainless steel wire brushes and specific cutting discs. In workshops where both carbon steel and stainless steel are used, it is highly recommended to store stainless steel separately and use dedicated tools to avoid cross-contamination.

✓ **Tack-welding:** Due to the low thermal conductivity of stainless steel, more tack welds are required to prevent deformation during welding. Follow this tack-welding sequence:

1. First tack weld at the far-right end
2. Second tack weld at the far-left end
3. Third and fourth tack welds at the opposite corners
4. Add additional tack welds every 1½″ to 1¾″ (3–4cm) apart

This distribution of tack welds helps to control deformation during welding.

2.   Protection against oxidation

✓ **Chromium oxides:** The reverse side of the weld is particularly sensitive to chromium oxide formation when exposed to air. When welding with SMAW, the slag generated can provide sufficient protection. However, in MAG or TIG welding, additional gas shielding is required on both sides.

3.   Temperature control and preheating

✓ **Chromium carbide formation:** Stainless steel is prone to chromium carbide precipitation in the heat-affected zone (HAZ), which reduces corrosion resistance. It is crucial to control the temperature during welding and, if necessary, apply controlled preheating to minimise this risk.

✓ **Electrode preheating:** For AWS 308-L electrodes, it is recommended to preheat them in an oven following the manufacturer's instructions. This improves performance by eliminating moisture from the coating, which reduces the risk of porosity in the weld.

4.   Self-releasing slag

✓ **The slag generated by 308-L electrodes is self-releasing,** meaning that it detaches easily after cooling. This property is due to the specific composition of the electrode coating, designed to prevent slag adhesion to the weld bead. However, when removing the slag, be cautious, as it can detach forcefully and cause burns. Always keep your welding helmet on when removing slag.

CONCLUSION

When welding AISI 304-L stainless steel with SMAW, the same basic techniques used for carbon steel can be applied, but adjustments must be made to account for the specific characteristics of stainless steel and the electrode. These include:

✓ Increased focus on cleanliness
✓ Preventing oxidation on the reverse side
✓ Precise temperature control
✓ Proper handling of self-releasing slag

By mastering these techniques and considerations, you will be well-prepared to produce high-quality stainless steel welds, preserving the corrosion-resistant properties of the material and ensuring the integrity of welded joints.

Wait . . . Too many technical details? No worries! Let's simplify it.

WELDING STAINLESS STEEL: WHERE DO I START?

Let's talk about how to weld stainless steel 304-L, which is one of the most commonly used types due to its great properties for many applications. The good news is that welding stainless steel is quite similar to welding carbon steel, but you need to pay extra attention to cleanliness and preparation.

Stainless steel is basically steel with added elements like chromium (12–30%), molybdenum, titanium, and nickel, making it more resistant to oxidation and corrosion, as well as more durable and easier to weld.

However, if not handled properly, it can lose its properties. Here's how to avoid mistakes:

✓ **Cutting and preparation:**
  Always use cutting discs specifically designed for stainless steel (labelled "inox"). If your workshop also works with carbon steel, store stainless steel separately and use dedicated tools for it. Carbon steel can "contaminate" stainless steel, which is something you want to avoid.

✓ **Brushing:**
  When brushing stainless steel, use only stainless steel wire brushes (not regular steel brushes used for carbon steel). Stainless steel brushes usually have silver-coloured bristles, so check them before starting.

✓ **Tack-welding:**
  Unlike carbon steel, you can't just use two tack welds and hope for the best. Since stainless steel doesn't conduct heat well, poor tacking can cause warping or excessive opening of the joint. I recommend at least five tack welds:
  • One at each end
  • One in the centre
  • Two more between the centre and the ends, ensuring no gap is greater than 1½″ (4cm)

✓ **Oxidation on the reverse side:**
  Stainless steel is very sensitive to oxidation on the back side of the weld when exposed to air. With SMAW electrodes, slag usually provides enough protection, but when using MAG or TIG, you'll need shielding gas protection on both sides.

✓ **Electrodes:**
  When using 308-L stainless steel electrodes, preheat them in an oven before welding, following the manufacturer's recommendations. This removes moisture from the coating, helping them perform better and preventing porosity in the weld.

Final tip: When removing slag, always keep your welding helmet on—slag can detach violently as it cools, and you don't want to get burned.

That's it! Now you know how to weld stainless steel properly, avoid common mistakes, and ensure top-quality welds!

WELDING DEFECT TERMINOLOGY

• **Excess weld reinforcement.** Excessive weld metal on the surface of the joint, either on the root or capping pass, usually due to excessive filler metal deposition.
• **Lack of penetration (LOP).** Insufficient fusion at the root of the weld. This defect occurs due to insufficient heat input, poor joint preparation (such as a narrow bevel, excessive land, or insufficient root gap), or contamination by oxides preventing full fusion. Excess filler metal can also prevent proper penetration.
• **Misalignment (High-low or joint misalignment).** The welded parts are offset, resulting in a height difference between them. TIG welding (GTAW) tends to cause more distortion due to its slower travel speed and high heat input, compared to MMA (SMAW) or MIG/MAG (GMAW). If the parts are not secured properly with backing bars or

strong tack welds, misalignment may occur. However, in most cases, it is a result of improper assembly.

- **Undercut.** A groove or depression along the weld toe or root, reducing the thickness of the base metal and weakening the joint. This is usually caused by poor technique, excessive current, which burns into the edges, or low heat input with excessive filler metal, causing the weld metal to build up in height instead of fusing properly with the base material.
  - o Dilution refers to the region where the base metal and filler metal mix. Poor dilution can increase undercutting risks.
- **Lack of fill (Undersized weld or insufficient throat thickness).** This occurs when too much heat input causes excessive concavity or when an undersized filler material (electrode, rod, or wire) is used. An undersized fillet weld or concave weld profile weakens the joint.
- **Asymmetrical weld bead.** The weld bead has a wider base compared to its height. This is common in fillet welds, where the weld throat is insufficient, leading to an imbalanced weld profile.
- **Cracking.** A critical defect due to its tendency to propagate and cause weld failure. Cracking is more likely to occur in medium/high-carbon steels and cast iron if preheating is not applied, or if the joint is excessively restrained, leading to thermal stresses that promote cracking.
  - o TIG crater cracks: At the end of a TIG weld, crater cracks can form due to rapid contraction when the arc is extinguished. This occurs if the down-slope function is not used, or if the weld pool is not properly filled at the end of the pass.
- **Porosity**: **Trapped gas bubbles within the weld metal,** forming small voids.
  - o Porosity occurs when gases become trapped in the weld pool due to:
    - Oxides or cutting fluid residues from poor joint cleaning.
    - Contamination in the weld pool (e.g., moisture, oil, grease).
    - Shielding gas failure (e.g., sudden torch movements dispersing the gas, low gas flow rate, or leaks in the shielding gas supply).
    - Low welding current, causing the weld pool to solidify too quickly before trapped gases can escape.

COMMON WELDING DEFECTS IN MANUAL METAL ARC (MMA) WELDING

1. Porosity

- **What is it?**
- Porosity appears as small cavities or gas pockets trapped within the weld metal or on the surface of the weld bead. These bubbles are formed by gases that become trapped in the molten pool before solidification.
- **Why does it occur?**
  - o The base material or electrode is contaminated with oil, grease, rust, or moisture.
  - o The electrode has absorbed moisture due to improper storage.
  - o Excessive travel speed prevents gases from escaping.
  - o A humid or draughty environment affects arc stability.
- **How to prevent it?**
  - o Thoroughly clean the base metal before welding.

o   Store electrodes in a dry place and, if necessary, re-bake them according to the manufacturer's recommendations.
o   Weld in a protected area, away from draughts and excessive humidity.
o   Maintain a consistent and appropriate travel speed.

2.   Undercut

- **What is it?**
- Undercut appears as a groove or depression at the toe of the weld bead where the base metal has been eroded but not adequately filled.
- **Why does it occur?**
   o   Excessive welding current, causing excessive arc penetration.
   o   Incorrect electrode movement (too fast or improper technique).
   o   Incorrect electrode angle during welding.
- **How to prevent it?**
   o   Adjust the current according to the material thickness and electrode type.
   o   Maintain a steady and uniform movement to ensure proper deposition of filler metal.
   o   Hold the electrode at the correct angle (typically between 5° and 15°, depending on the welding position).

3.   Slag inclusions

- **What is it?**
- Slag inclusions are non-metallic particles trapped within the weld metal, which weaken the joint and can lead to failure under stress.
- **Why does it occur?**
   o   Inadequate slag removal between passes.
   o   Excessively slow travel speed, allowing slag to become embedded in the molten pool.
   o   Incorrect electrode angle, preventing proper slag ejection.
- **How to prevent it?**
   o   Clean each pass thoroughly before applying the next layer.
   o   Adjust travel speed to prevent slag entrapment.
   o   Maintain the correct electrode angle to allow slag to flow out of the weld bead.

4.   Lack of fusion

- **What is it?**
- Lack of fusion occurs when the weld metal does not properly bond with the base material or previous passes, creating a weak weld joint.
- **Why does it occur?**
   o   Insufficient welding current, leading to a weak arc.
   o   Excessive travel speed.
   o   Incorrect positioning of the electrode relative to the joint.
- **How to prevent it?**
   o   Increase the current to ensure adequate fusion.
   o   Reduce travel speed to allow sufficient time for fusion.
   o   Position the electrode correctly for optimal penetration.

PRACTICAL SOLUTIONS TO PREVENT AND CORRECT WELDING DEFECTS

1. **Welding parameter adjustments**
   - **Current settings:** Adjust the amperage according to the electrode diameter and base metal thickness. High current can cause undercut, while low current can lead to lack of fusion.
   - Travel speed: Maintain a steady pace to ensure a well-formed weld bead.
   - **Arc length:** Keep a constant distance between the electrode and the weld pool (approximately equal to the electrode diameter).
2. **Welding technique**
   - **Electrode angle:** Maintain the correct inclination, typically **slightly tilted in the direction of travel**.
   - **Electrode movement:** Use a controlled technique (zigzag, circular, or U pattern) depending on the joint type to ensure even filler metal distribution.
   - **Cleaning between passes:** Remove slag completely with a chipping hammer and wire brush before proceeding with additional weld passes.
3. **Environmental control**
   - **Protect the welding area from draughts and moisture** using welding screens or barriers.
   - **Preheat the base metal** if working in cold conditions or on thick materials to avoid cracking and lack of fusion.
4. **Equipment maintenance**
   - **Inspect electrode holders and clamps** to ensure they are clean and in good condition.
   - **Use appropriate cables** and check electrical connections to prevent voltage drops affecting arc stability.

CONCLUSION

Identifying and preventing defects in **MMA (SMAW) welding** is essential to achieving strong and durable welds. By adjusting parameters correctly, applying the right technique, and ensuring attention to detail, **welding defects can be minimised**, resulting in high-quality welds. Continuous practice and learning are key to refining your skills as a welder.

2 TIG Welding

TEAMWORK "THE INVISIBLE ROPE"

A group of monks were trying to move a large rock that was blocking the path to their monastery. After many failed attempts, one of them said, "We will never move it on our own." Then, they all joined hands around the rock and began to pull together, as if an invisible rope connected them. The rock, which had seemed immovable, finally gave way.

When we work together, even the impossible becomes achievable.

INTRODUCTION TO TIG WELDING

I will keep this brief, just like in the previous chapter, and summarise the most important things you need to know about the Tungsten Inert Gas (TIG) welding process.

THE HISTORY OF TIG WELDING

The 1930s were a revolutionary period for welding. By then, the technology to store industrial gases in bottles had already been developed, which led to the rise of oxyacetylene welding—a process that uses a flame created by mixing oxygen and acetylene to weld thin materials. At the same time, shielded metal arc welding (SMAW) was becoming widely adopted in the industry. Although it was the most popular process at the time, it had its limitations—for example, it was not suitable for welding highly reactive materials, such as titanium.

In 1940, researchers discovered a new application of the electric arc that could perform such tasks with superior quality: TIG welding. This process is essentially the electrical version of oxy-fuel welding. It works by passing an electric current through a non-consumable electrode, which heats up without melting. The heat generated is used to fuse the base metal, and if required, filler metal can be added in the form of a rod, which the welder controls with their other hand—similar to how it is done in oxyacetylene welding. Once the weld pool cools, the weld bead is formed.

The electrode in the TIG process is made of tungsten (also known as wolfram), which has the highest melting point of any metal. This electrode can be pure tungsten or alloyed with small amounts of other elements (typically no more than 5%).

WELD PROTECTION

Unlike SMAW, in the TIG process, neither the tungsten electrode nor the filler rod has a protective coating. Instead, this process uses a shielding gas system similar to MIG/MAG welding. The gas—usually argon or helium—is supplied through the torch, displacing the surrounding air and creating an inert atmosphere that protects the weld pool.

ADVANTAGES OF THE TIG PROCESS

TIG welding is still considered the highest-quality manual welding process today, although it is somewhat slower compared to wire-fed or stick welding. Its advantages are numerous:

DOI: 10.1201/9781003422488-3

- It produces no slag or spatter.
- It can be used to weld most metals.
- It is not excessively expensive.
- It is suitable for welding in all positions.
- It is widely used for welding root passes in pipe welding.
- The welds obtained are of high quality and purity.

For these reasons, TIG welding is widely used for critical applications in industries such as shipbuilding, aerospace, nuclear, chemical, space exploration, and the oil and gas sector.

FUNDAMENTALS OF TIG WELDING

The necessary equipment (see Figure 2.1):

- **Power source (rectifier-transformer):** Similar to those used in SMAW welding, it can be AC or DC. It may also include a high-frequency (HF) generator to stabilise the arc and facilitate ignition without touching the workpiece with the tungsten electrode.
- **Gas circuit:** The shielding gas is supplied from a bottle through a pressure regulator, ensuring the necessary protection for the weld.
- **Torch:** Holds the tungsten electrode and directs the shielding gas.
- **Torch cooling system:** Often referred to as *water cooling*, though it actually uses a non-conductive coolant (never replace it with water or automotive antifreeze). It is not essential but is recommended when working at high amperages.
- **Solenoid valves and regulators:** Typically integrated into the welding machine, synchronising the cooling system (if present), gas flow, and current.
- **Auxiliary equipment:** Such as foot pedals to control amperage.

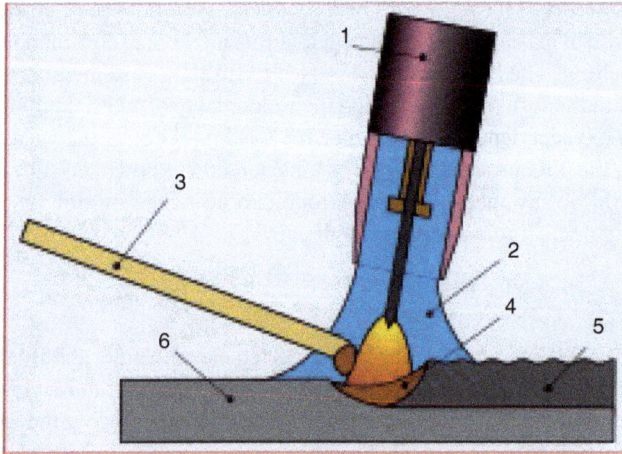

1. Gas nozzle   2. Gas shielding   3. Filler rod   4. Weld pool
5. Weld bead   6. Base metal

FIGURE 2.1 Diagram of the TIG welding process. 1. Gas nozzle; 2.Shielding gas; 3.Filler rod; 4. Weld pool; 5.Weld bead; 6.Base metal.

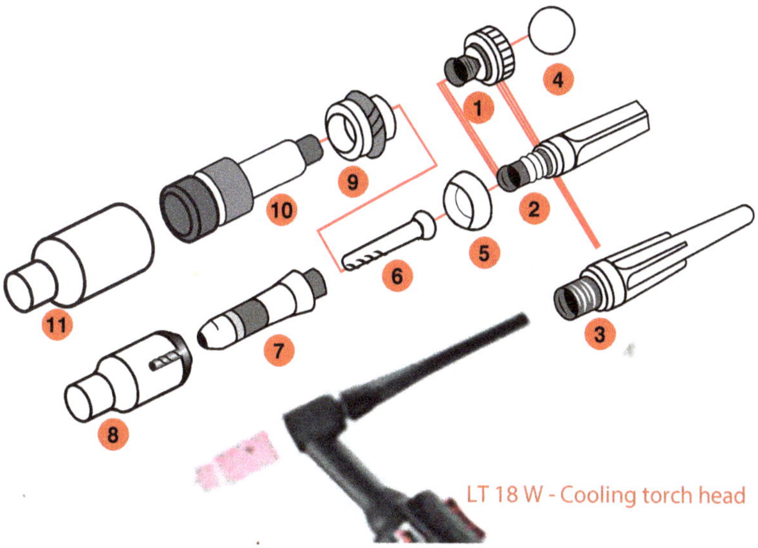

LT 18 W - Cooling torch head

FIGURE 2.2 Exploded view of a TIG welding torch model LT 18 W with a cylindrical head. Numbered components are shown: 1, 2, and 3: Back cap. 6; Collet.7 and 10; Gas diffuser. 8 and 11 Ceramic nozzle.

TORCH COMPONENTS

A basic TIG torch consists of the following parts (see Figure 2.2):

- **Nozzle:** Usually made of ceramic, this is the nozzle through which the shielding gas flows. The diameter varies depending on the tungsten size and the type of weld being performed. Nozzles are fragile, especially when hot.
- **Collet:** Sometimes made of copper, its function is to grip the tungsten electrode, preventing it from slipping.
- **Collet body (gas diffuser):** Holds the collet and is typically attached to the torch via a delicate thread (should not be removed while hot). It also distributes the shielding gas evenly.
- **Back cap:** The rear part of the torch body, which is removed to insert or release the tungsten electrode.

NON-CONSUMABLE TUNGSTEN ELECTRODES AND THEIR TYPES

The role of tungsten in TIG welding is to generate the necessary heat when the arc strikes between the electrode and the workpiece, allowing a current to pass through. Different types of tungsten electrodes are used depending on the material being welded. Here are the most common types:

Pure tungsten (WP—Green)

- **Composition:** 99.5% pure tungsten, with no alloying elements.
- **Properties:** High electron emission capacity, primarily suited for alternating current (AC). Provides a stable arc but has lower durability and is more prone to contamination than alloyed tungsten electrodes.

- **Applications:** Used mainly for aluminium and magnesium welding in AC.
- **Health:** Non-radioactive, making it safe to use.

Thoriated tungsten (WTh20 red)

- **Composition:** 98% tungsten, 2% thorium (ThO_2).
- **Properties:** Excellent arc stability in direct current (DC), with greater durability than pure tungsten. Ideal for high-amperage applications.
- **Applications:** Best for welding carbon steels, stainless steels, and nickel alloys in DC.
- **Health:** Radioactive. Protective measures are recommended to avoid inhaling radioactive dust when grinding electrodes.

Ceriated tungsten (WCe20—Grey)

- **Composition:** 98% tungsten, 2% cerium (CeO_2).
- **Properties:** Easy arc starting and stable performance at low current in DC. Less durable than thoriated tungsten but non-radioactive.
- **Applications:** Suitable for precision work and thin sheets in DC.
- **Health:** Non-radioactive, making it safe for regular use.

Lanthanated tungsten (WLa15/20—Black/Gold)

- **Composition:** 98% tungsten, 1.5–2% lanthanum (La_2O_3).
- **Properties:** Excellent arc starting and stability, with a longer lifespan than ceriated tungsten. Versatile for DC welding.
- **Applications:** Used for welding steel and ferrous alloys.
- **Health:** Non-radioactive, making it safe for use.

Zirconiated tungsten (WZr08—White)

- **Composition:** 98% tungsten, 0.15–0.4% zirconium (ZrO_2).
- **Properties:** Excellent arc stability and resistance to contamination, ideal for AC welding. Less prone to contamination and cracking.
- **Applications:** Primarily used for aluminium and magnesium welding in AC.
- **Health:** Non-radioactive, making it safe for use.

Rare-earth tungsten (WTL/E3—Purple)

- **Composition:** A blend of rare-earth oxides, such as lanthanum, cerium, and yttrium, combined with tungsten.
- **Properties:** Combines the best characteristics of lanthanated and ceriated tungsten, offering excellent arc ignition, stability, and durability. Known for having a lower erosion rate and a long lifespan.
- **Applications:** Versatile, used in both AC and DC welding, ideal for steel, aluminium, and other non-ferrous alloys.
- **Health:** Non-radioactive, making it safe for use.

Applications and specific materials

- **Pure tungsten:** Best for AC, used in aluminium and magnesium welding.
- **Thoriated tungsten:** Ideal for DC and high-current applications in carbon steel, stainless steel, and nickel alloys.

- **Ceriated and lanthanated tungsten:** More versatile, primarily used in DC. Preferred for precision work and welding of light metals.
- **Zirconiated tungsten:** Best for AC welding of aluminium and magnesium due to its excellent resistance to contamination.
- **Rare-earth tungsten:** Versatile, long-lasting, and can be used in both AC and DC welding.

Price Differences

- **Pure tungsten:** Generally the least expensive.
- **Thoriated tungsten:** Affordable, but its use is declining due to health concerns.
- **Ceriated and lanthanated tungsten:** Slightly more expensive but offer greater safety and versatility.
- **Zirconiated and rare-earth tungsten:** Typically more costly due to their advanced properties and versatility.

Health and safety considerations

- **Thoriated tungsten:** The only radioactive option. Protective respiratory equipment and proper ventilation should be used when grinding these electrodes.
- **Other tungsten electrodes:** Non-radioactive, making them safer for regular use.

Tungsten electrode grinding

All tungsten electrodes must be ground to a fine point to focus the arc heat on a small area, ensuring a well-defined weld pool (see Table 2.1).

CONCLUSION

Each type of tungsten has its ideal applications and advantages. The choice depends on the base material, type of current, and safety considerations. While thoriated tungsten offers excellent performance in DC welding, its radioactivity makes it a less desirable option today. Tungsten alloys with cerium, lanthanum, zirconium, and rare earth elements provide comparable performance with superior safety, making them more popular in modern industry.

FILLER RODS

In TIG welding, the torch provides the necessary heat to melt the base metal while supplying shielding gas, whereas the filler material is manually added using filler rods. These rods must have a chemical composition similar to the base metal and should always be clean. The available diameters are 1; 1.6; 2; 2.4; 3.2; 4, and 4.8mm, with a typical length of approximately 900mm.

TABLE 2.1

Diámetro de tungstenos mas utilizados (mm)	Altura máxima del cono (mm)	Altura mínima del cono (mm)
1	2	1,5
1,6	3,2	2,4
2	4	3
2,4	4,8	3,6
3,2	6,4	4,8
4	8	6

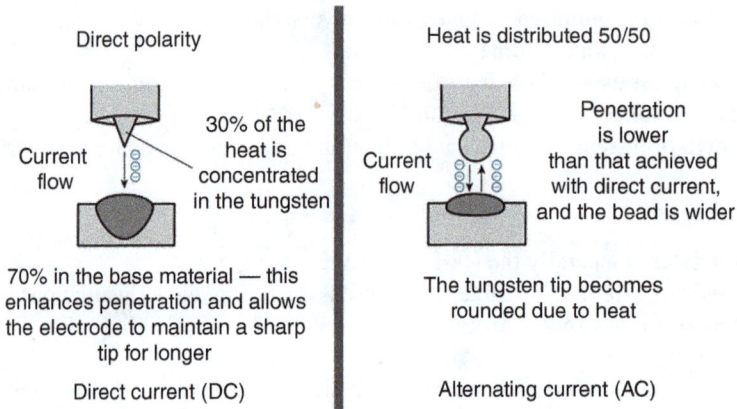

Direct polarity

Heat is distributed 50/50

Current flow

30% of the heat is concentrated in the tungsten

Current flow

Penetration is lower than that achieved with direct current, and the bead is wider

70% in the base material — this enhances penetration and allows the electrode to maintain a sharp tip for longer

The tungsten tip becomes rounded due to heat

Direct current (DC)

Alternating current (AC)

FIGURE 2.3   Comparison between TIG welding with direct current (DC) and alternating current (AC). On the left, DCEN concentrates 70% of the heat on the base material, enhancing penetration and keeping the tungsten tip sharp. On the right, AC distributes heat 50/50, reducing penetration and rounding the tungsten tip.

SELECTING THE TYPE OF CURRENT (SEE FIGURE 2.3)

- **Direct current (DC):** Primarily used in Direct Current Electrode Negative (DCEN), concentrating heat on the workpiece and preserving the tungsten tip. In Direct Current Electrode Positive (DCEP), the tungsten overheats and deteriorates quickly.
- **Alternating current (AC):** Essential for welding aluminium and magnesium due to its oxide-cleaning effect. Although it provides less penetration and a less stable arc, it is necessary for these materials.

SHIELDING GASES

- **Argon:** The most commonly used gas due to its high density and low ionisation energy, which facilitates arc ignition and stability. However, its low thermal conductivity limits penetration.
- **Helium:** Provides better penetration and a wider weld bead profile, but is more expensive and less stable than argon. It is often mixed with argon for welding materials such as aluminium.

GAS REGULATOR

The gas regulator is essential for supplying shielding gas to the welding equipment, allowing the welder to set the correct working pressure (see Figure 2.4).

A standard regulator consists of two pressure gauges:

- The **high-pressure gauge** indicates the gas level in the cylinder when opened.
- The **low-pressure gauge** allows the user to set the working pressure.

Parts of the pressure regulator and marking descriptions

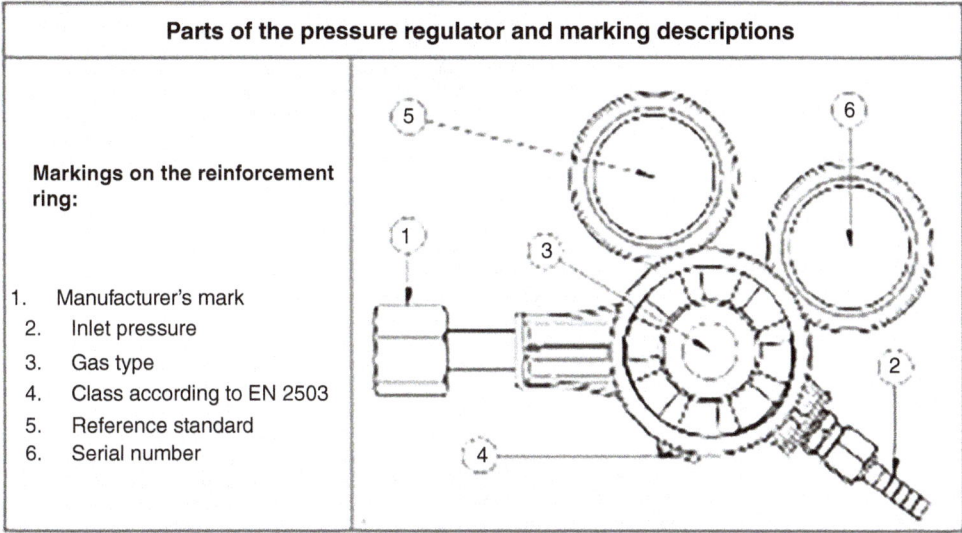

Markings on the reinforcement ring:

1. Manufacturer's mark
2. Inlet pressure
3. Gas type
4. Class according to EN 2503
5. Reference standard
6. Serial number

FIGURE 2.4   Diagram of a pressure regulator with a description of its markings.

1. Cylinder connection; 2. Hose outlet; 3. Pressure adjustment screw; 4. Safety valve; 5.High-pressure gauge; 6.Low-pressure gauge (Provided courtesy of GALAGAR).

It has a delicate adjustment mechanism (a screw beneath the gauges), which should be fully loosened (turned to the left) before opening the cylinder. Once the cylinder is open, turning the regulator to the right increases the reading on the low-pressure gauge. If excessive pressure is set, it must be reduced by loosening the screw, purging the hose, and restarting the adjustment.

The regulator has a threaded connection designed for a specific type of cylinder. If the fitting does not screw in easily, **never lubricate or force the threads**.

AUXILIARY WELDING FUNCTIONS (SEE FIGURE 2.5)

- **"Two-stroke" and "four-stroke" modes:** Allow efficient control of gas flow and welding current.
- **Pre-flow and post-flow of shielding gas:** Ensure the weld is properly shielded before and after welding.
- **Current upslope and downslope:** Provide a smooth start and finish, preventing craters and improving weld quality.
- **"Lift arc" ignition:** Provides precise arc initiation by gently scraping the tungsten against the workpiece, similar to a stick electrode. This avoids the use of high frequency where it may cause interference (e.g. with computer equipment or pacemakers).
- **Pulsed arc:** Reduces heat input, making it ideal for heat-sensitive materials.
- **Remote control:** The most popular and user-friendly option is the **foot pedal**, which allows the welder to adjust the current during welding for greater comfort and control (see Figure 2.6).

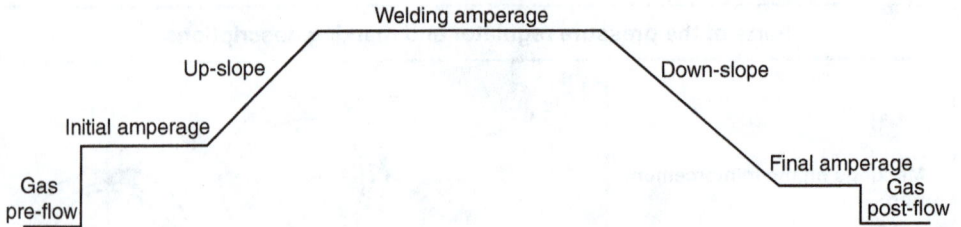

FIGURE 2.5   Graph showing the progression of current in a welding cycle. It represents the stages of pre-flow, gradual amperage increase, welding amperage, progressive decrease, and post-flow.

Courtesy of MILLER.

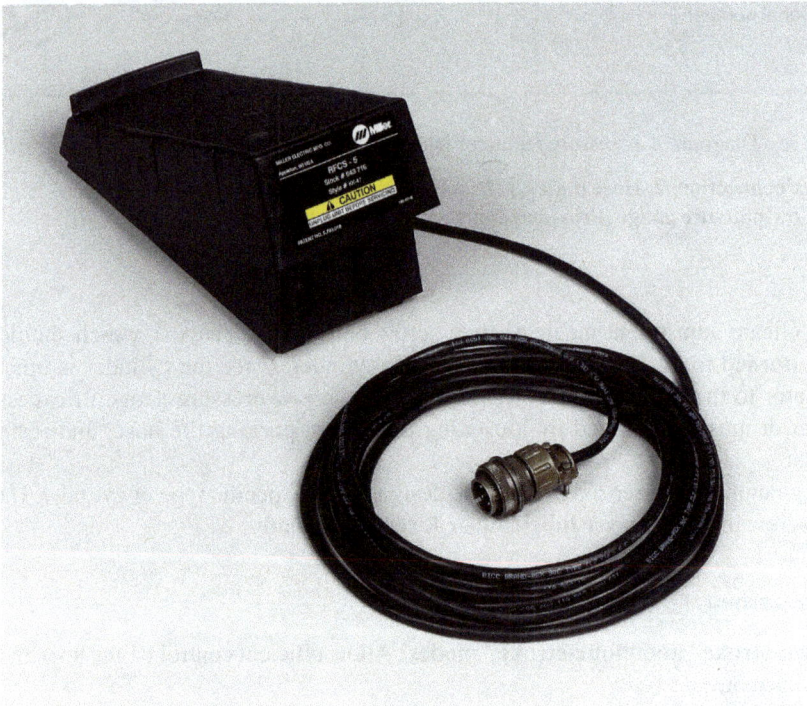

FIGURE 2.6   Remote foot control for TIG welding. Device used to regulate welding current via foot pressure, with a cable and connector for power source connection. Courtesy of ITW (MILLER).

DIFFERENCES BETWEEN THE STANDARD NOZZLE AND GAS LENS IN TIG WELDING

In the TIG welding process, the choice of nozzle is key to controlling arc protection and the weld pool through the flow of inert gas, typically argon. There are two main types of nozzles commonly used: the **standard nozzle** and the **gas lens**. In the following section, we explain their differences, advantages, and applications.

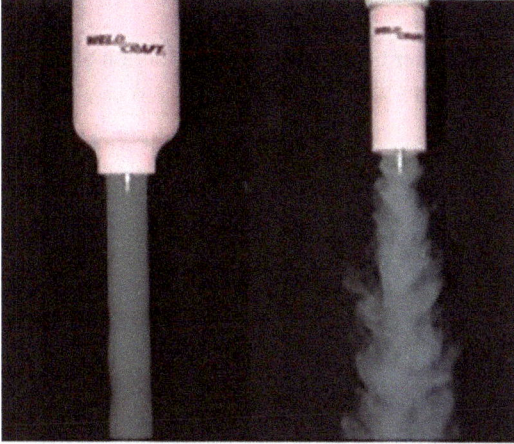

FIGURE 2.7   Comparison of shielding gas flow in a TIG torch. On the left, proper laminar flow; on the right, unstable turbulent flow that can affect weld quality. Courtesy of ITW (MILLER).

STANDARD NOZZLE

The **standard nozzle** is the most commonly used type in TIG welding. It is designed to direct the shielding gas flow uniformly over the arc and weld pool, providing adequate protection against contamination from the air (see Figure 2.7).

- **Design:** The standard nozzle has a simple structure, with an opening at its tip that allows the passage of inert gas. The gas flow is less concentrated than in a gas lens, but it is sufficient for most general applications.
- **Numbering system:** Nozzle sizes are based on a numbering system where the assigned number corresponds to the internal diameter divided by **1.6**. This means you can estimate the approximate internal diameter by multiplying the nozzle number by **1.6**.

Here are some examples:

- **Nozzle #4:** Approx. internal diameter = **4 × 1.6 = 6.4 mm ≈ 0.252 in**
- **Nozzle #5:** Approx. internal diameter = **5 × 1.6 = 8 mm ≈ 0.315 in**
- **Nozzle #6:** Approx. internal diameter = **6 × 1.6 = 9.6 mm ≈ 0.378 in**
- **Nozzle #7:** Approx. internal diameter = **7 × 1.6 = 11.2 mm ≈ 0.441 in**
- **Nozzle #8:** Approx. internal diameter = **8 × 1.6 = 12.8 mm ≈ 0.504 in**

These measurements are approximate, as the internal diameter may vary slightly depending on the manufacturer. However, this system provides a quick way to select the correct nozzle size without manually measuring each one. This selection process is essential for determining the required gas flow, as we will see in the practical welding exercises.

- **Advantages:**
  - o   More affordable and widely available.
  - o   Suitable for most standard TIG welding applications.

GAS LENS

The gas lens is a more advanced option that provides better control over the distribution of shielding gas. It features a fine metal mesh that creates a laminar gas flow, meaning the gas moves more smoothly and evenly, reducing turbulence.

- **Design:** The metal mesh inside the gas lens filters and distributes the gas more evenly, creating a "curtain" of shielding gas that is more stable and efficient. This is especially useful when greater gas coverage is needed or when welding materials that are highly sensitive to oxidation.
- **Numbering system:** Similar to the standard nozzle, gas lens sizes range from #4 to #8, though larger sizes such as #10 are sometimes used for special applications.
- **Advantages:**
  o Provides wider gas coverage with lower gas flow rates.
  o Improves weld quality, particularly in difficult positions or when working with oxidation-sensitive materials.
  o Allows the tungsten to extend further beyond the nozzle without compromising gas protection, improving visibility and access to hard-to-reach areas.

CONCLUSION

Both systems have advantages and are suitable for different TIG welding applications. If you're **just starting out** and working on standard welds, the **standard nozzle** is more than sufficient. However, if you require **greater gas control** or are working on **critical applications**, you should try the **gas lens** to see how it improves your technique and weld quality.

Ultimately, the **best choice depends on your specific needs**, so we encourage you to experiment with both types of nozzles to find the one that best suits your work.

IT'S NORMAL TO FEEL LIKE YOU'RE NOT IN CONTROL AT FIRST.

TIG welding, with its precision, can be intimidating. However, think of it as a dance between your hands and the material: the more you practise, the more harmony you will achieve. If you feel frustrated, remember that even the best welders were once in your position. Breathe, adjust your posture, and try again. **You will get there.**

Very important: Best practices in the welding workshop

- **Protection:** Always wear safety glasses when grinding the tungsten electrode. When welding carbon steel, the cone should always be twice the electrode diameter (e.g. 1.6mm tungsten—3.2mm cone; 2.4mm tungsten—4.8mm cone).
- **Dismantling the torch: NEVER** dismantle a hot torch, as this could damage the threads of its components. Allow it to cool to prevent burns and maintain the equipment in good condition.
- **Handling the torch:** Place the torch carefully on the workbench. Ceramic nozzles are fragile, especially when hot.

- **Tungsten maintenance:** Always clean the tungsten on the grinder if it accidentally touches the weld pool, the workpiece, the filler rod, or if it becomes rounded. A clean cone concentrates the arc in a small, easy-to-control point; a dirty one does the opposite.
- **Torch distance:** Control the distance between the torch and the workpiece to ensure the shielding gas does its job. The recommended **gap is no more than 3–4mm.**
- **Gas flow and tool selection:** Use an appropriate gas flow rate for the nozzle size, a nozzle suitable for the type of work (narrow nozzles for penetration passes, wider ones as the weld bead increases), and a filler rod that best matches the task (thinner for penetration or small thicknesses, thicker for capping passes or wide plates).
- **Cleaning filler rods:** It is good practice to wipe down the filler rods before using them; otherwise, all the dirt will be transferred into the weld. Bend the last 3cm of the rod tip (like the handle of an umbrella) to prevent accidents.
- **Weld bead tie-ins:** Practise tying in weld beads in all exercises (to avoid porosity, allow the last bead to cool before continuing).
- **Using advanced functions:** Make use of high-frequency start and pulsed arc whenever your welding machine allows it, as well as current upslope and downslope, preflow and post-flow of shielding gas, and the four-stroke mode (which is much more comfortable).
- **Weld bead quality:** Practise until your weld beads look like true works of art—that's what is expected of you. With TIG, weld beads should stand out.
- **Polarity:** Always use **DCEN** when welding carbon steel and stainless steel.
- **Handling hot workpieces:** Use pliers or tongs to pick up hot workpieces, never your hands—even with welding gloves! Workpieces can be extremely hot.
- **Workshop safety:** Always consult your instructor if you have any doubts about how to use any tool, especially grinders, saws, or bench grinders. These machines require respect and proper safety precautions, though never fear.
- **Personal protection: NEVER** look at the arc without a welding helmet. **NEVER** weld without gloves or wearing short sleeves. Always wear protective gear such as aprons, sleeves, and spats, and return them to their place after use. Wearing a mask is also advisable.
- **Caring for equipment:** Be careful with all workshop equipment and help keep your workstation clean at the end of each session.

TIG WELDING ON CARBON STEEL

PRACTICE 1. FIRST BEADS IN FLAT POSITION

- **Base material:** Carbon steel plate, 4" x 4" x 1/8".
- **Electrodes:** Choose from available options: Ø 1/16" tungsten, either thoriated (WTh20 red), lanthanated (WLa10/15 black or gold), ceriated (WCe20 grey), or rare-earth (E3 purple).
- **Nozzle:** No.5 or No.6 (Ø 5/16" and Ø 3/8", respectively).
- **Filler rod:** Ø 1/16" or 5/64", type ER 70S-6.
- **Number of beads:** Nine straight beads with no oscillation.
- **Current:** 55–75 Amps.
- **Gas flow rate:** 17–21 cubic feet per hour (CFH) (1 CFH per 1/16" of nozzle diameter).
- **Auxiliary tools:** Ruler, scriber, tape measure, file, grinder, centre punch, and hammer.

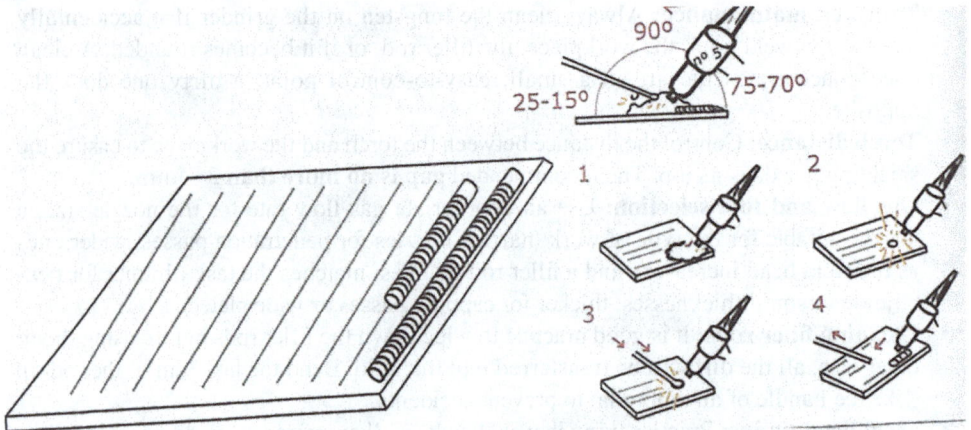

FIGURE 2.8 TIG welding practice on a metal plate. On the right, diagrams show the torch inclination and the sequence for arc ignition, bead initiation, and controlled travel.

WELCOME TO YOUR FIRST EXPERIENCE IN TIG WELDING!

In this exercise, we will carry out our first straight beads in the flat position.

To begin, we will measure the plate using a ruler or tape measure. It is essential that, when cutting the material, the dimensions match those specified in the practice. To ensure accuracy, we will make scribing marks with a scriber and then highlight them using a centre punch, marking where the cut should be made.

Once the piece has been cut, it is necessary to clean its surface. We will use a grinder with a sanding disc to achieve a smooth finish, removing oxides, grease, cutting fluid, etc. (or a grinding disc if no sanding disc is available, taking care not to reduce the thickness too much). Next, we will deburr the sharp edges of the cut and round off the four corners using a grinder or hand file.

After polishing the piece, we will mark each 3/8″ along one edge of the square and repeat the process on the opposite edge, marking a total of nine lines. Using a ruler and scriber, we will connect each mark to its counterpart on the other side, creating nine straight lines spaced 3/8″ apart, which will serve as a reference for welding the beads.

Finally, we will use a centre punch and hammer to create small marks along the lines, helping to follow the weld path more accurately.

With these steps completed, the piece will be ready to weld. Now, we will sharpen the tungsten (the cone should measure around 1/8″ from base to tip, approximately twice its diameter), adjust the gas flow rate, select the working mode (touch start, Lift-Arc, two-stroke, or four-stroke), set the welding current to the minimum value recommended, and put on the necessary protective equipment, leaving the welding helmet for the last moment (Figure 2.8).

READY? LET'S BEGIN

Position the torch at the right end of the plate. Right-handed welders are advised to weld from right to left (left-handed welders should do the opposite) and to position their head at the left end. It is crucial to "be at the end" of the weld bead to have a clear and constant view of the tungsten tip, ensuring a uniform arc length.

Although it may be uncomfortable at first, lowering your head to eye level with the torch will help you maintain full control of the weld. This observation technique will be constant throughout all exercises, so take a few seconds to find the best angle to monitor the weld.

The torch should be held at a 10° angle, tilted backwards in the direction of travel, but never tilted sideways, as this could compromise the shielding gas coverage over the weld.

STARTING THE WELD

When initiating the arc, keep the torch steady (maintaining a 1/8″ gap between the tungsten and the plate) until a small, flat weld pool forms (between 1/8″ and 3/16″ in diameter), ensuring it is fully molten. Then, move forward a few millimetres and repeat the process, forming the weld pool at each step until the bead is complete.

The objective is to create uniform beads, with clear ripples on their surface, indicating each controlled pause made during welding.

If you feel the bead is too dark or struggle to see the scribed line, don't worry—it's normal! If you lose the reference line, stop! Otherwise, you may stray from the intended path.

Take your time to achieve consistent bead width. Familiarise yourself with the torch before increasing the amperage.

INTRODUCING THE FILLER ROD

The filler rod should be used in sync with the welding process, dipping in and out of the weld pool in a controlled manner. Initially, a low filler addition rate can simplify the process, allowing the weld pool to remain completely liquid and flat (you'll notice a slight agitation on the surface due to gas pressure).

Don't rush—only dip the rod when the pool is fully ready. Lightly touch the filler tip to the molten pool, withdraw it a few millimetres, and wait for the pool to be ready for the next addition.

The result should be a slightly raised bead, not concave like the first welds without filler.

"The one who controls others is strong; the one who controls themselves is powerful."
    (Tibetan proverb)
    Each bead you weld reflects your dedication and attention to detail. Don't aim for perfection on your first try—instead, focus on progress and continuous improvement.
    Remember why you started and find within yourself the strength to keep going.

TIG WELDING ON CARBON STEEL

PRACTICE 2: HORIZONTAL FILLET WELD. PB (2F)

- **Base material:** Carbon steel plate, 6″ x 1.2″ x 0.3″ or 0.12″ thick.
- **Electrodes:** Select from available options: Ø 3/32" thorium (WTh20 red), lanthanum (WLa 10/15 black or gold), cerium (WCe 20 grey), or rare earth (E3 purple).
- **Nozzle:** No. 5 or No. 6 (5/16″ and 3/8″ diameter, respectively).
- **Filler rod:** Ø 1/16″ or 5/64″, type ER 70S-6.
- **Number of weld beads:** One bead, introducing the Walking the Cup technique.
- **Current:** 120–150 amps.
- **Shielding gas flow rate:** 17–21 cubic feet per hour (CFH) (1 CFH per 0.04″ of nozzle internal diameter).
- **Tungsten stick-out:** 1/8–5/32″.

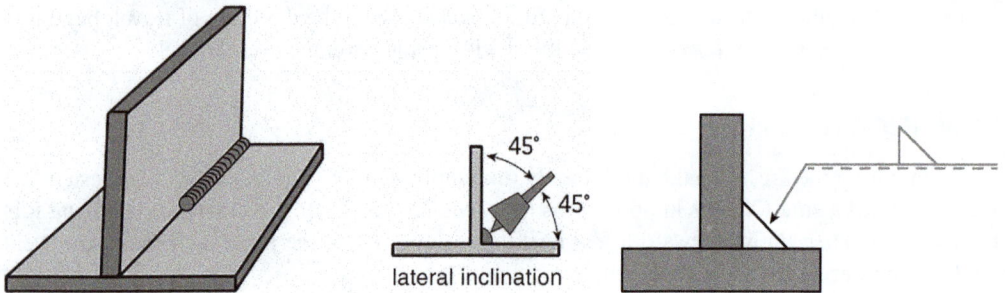

FIGURE 2.9   TIG welding in a fillet joint. On the right, a diagram shows the torch lateral inclination at 45° for proper material fusion.

In this practice, we will begin working with fillet welds (Figure 2.9).

1.   Initial preparation:

- **Clamping the plates:** Tack-weld the plates at 90° and secure them firmly to the workbench using, for example, a clamp. This is essential for properly applying the Walking the Cup technique, which requires the ceramic nozzle of the torch to be in contact with both the upper and lower plates of the joint.
- **Cleaning and material preparation:** Ensure the tungsten electrode is clean and properly sharpened. Wipe down the filler rod with a cloth to remove any dirt or contaminants that could affect weld quality.

2.   Torch positioning:

- **Angle and orientation:** Position the torch at 45° relative to both plates, with a slight backward tilt. This allows for clear visibility of the tungsten and prevents it from accidentally dipping into the weld pool.
- **Maintaining the angle:** Keeping a consistent torch inclination throughout the weld is crucial. This helps to prevent undercut at the top of the weld bead, a critical defect in welder qualification tests.

3.   Starting the weld bead:

- **Initial positioning:** Start from the right-hand side of the plate (if you are right-handed) or from the left-hand side (if left-handed). Ensure the ceramic nozzle is simultaneously in contact with both plates before initiating the weld.

4.   *Walking the Cup* technique:

- **Torch positioning and pressure:** Hold the torch firmly, ensuring the torch body is positioned at 45° relative to both plates. The ceramic nozzle must remain in continuous contact with both surfaces of the fillet joint. This contact provides a stable support, allowing better control over torch movement.
- **Side-to-side torch movement:** The key movement in this technique is a side-to-side rocking motion of the torch handle, from left to right and vice versa. This movement must be smooth and consistent. Its purpose is to allow the ceramic nozzle to roll forward along the joint without the tungsten electrode moving closer or farther from the

workpiece. It is important to remember that the sideways motion of the torch handle generates the forward movement, rather than an intentional push.

- **Torch advancement:** Forward motion results from the oscillating movement combined with slight pressure against the joint.
- **Coordination of movement:** With practice, you will develop a natural feel for combining the lateral oscillation with forward movement. This should be a fluid and continuous motion, where the side-to-side movement helps "walk" the nozzle along the plate surface, while gentle forward pressure ensures steady progress.

5.   Dry practice:

- **Exercise with the machine off:** Before starting to weld, practise the movement dry, without striking an arc. Focus on getting comfortable with the rocking motion of the torch handle and the forward movement of the ceramic nozzle along the joint. This will help build confidence and refine control before introducing the arc and weld pool.

6.   Welding exercise (without filler metal):

- **Fusion-only weld:** Once comfortable with the motion, carry out the exercise by welding without adding filler metal. Focus on the weld pool, ensuring it remains consistent in size and that the Walking the Cup technique is applied smoothly and evenly.

7.   Welding with filler metal:

- **Introducing the filler rod:** Repeat the exercise, but this time add filler metal. Apply what you learned in the first practice regarding coordination between the torch and the filler rod. Remember, the filler rod should be introduced at the edge of the weld pool, not at its centre, to prevent unnecessary cooling.

8.   Practice on thinner plates:

- **Adapting to different thicknesses:** Perform a new welding attempt on a test piece of the same dimensions but with a thickness of 0.12″ or even 0.06″. This will help you adapt the technique to different material thicknesses and refine control over both the torch and the filler rod.

HOLDING AND HANDLING THE FILLER ROD IN TIG WELDING

Precise control of the filler rod is essential for achieving high-quality weld beads in TIG welding. In the following section, I explain how to hold the filler rod and use your fingers to advance the material towards the weld pool effectively.

1.   Holding the filler rod

The way you hold the filler rod directly impacts your ability to control it during the welding process. Here are some guidelines:

- **Hand position:** Hold the filler rod between your thumb, index finger, and middle finger of your non-dominant hand (left hand if you are right-handed, right hand if you are left-handed). These three fingers are mainly responsible for keeping the filler rod stable and under control.
- **Firm but relaxed grip:** The filler rod should be held firmly, but without excessive pressure. A too-tight grip may cause hand fatigue and make it harder to apply the fine

control required for precise filler metal application. Keep your fingers relaxed and ready to move smoothly.

- **Distance from the tip:** Leave a sufficient distance between the tip of the filler rod and your hand (4″ to 6″), allowing for proper control while advancing the rod towards the weld pool.

2.   Using your fingers to advance the rod

The movement of advancing the filler rod into the weld pool should be smooth and controlled. Here's how to use your fingers to achieve this:

- **Thumb, index, and middle finger:** To advance the filler rod, use your index and middle fingers as a support point and slide the rod forward with your thumb. This movement should be small and controlled, allowing you to push the rod into the weld pool with precision.
- **Small and controlled movements:** Instead of making large and abrupt movements, push the filler rod forward in small increments. This will help maintain consistent control over the amount of material being introduced into the weld pool.
- **Readjusting the grip:** As the filler rod melts, you will need to readjust your grip. To do this without losing control, slide the rod forward with your fingers while maintaining a firm grip. You can do this quickly when the weld pool is stable and does not require immediate filler metal addition.

3.   Additional techniques for precise filler metal application.

- **Continuous sliding:** Some welders prefer to keep the filler rod in continuous motion along their fingers, advancing the rod smoothly while welding. This is particularly useful for long welds, where constant filler application is required.
- **Rod rolling technique:** Another method is to gently roll the filler rod between your thumb and fingers while advancing it. This movement allows for even finer control, as you can adjust the feed rate without releasing the rod.
- **Avoid abrupt movements:** Avoid moving the rod too quickly or abruptly into the weld pool, as this may cause excessive filler application and unnecessary cooling of the pool. Always keep movements smooth and calculated.

4.   Hang and rod positioning

- **Rod inclination:** It is advisable to incline the filler rod at an angle of 15° to 30° relative to the workpiece. This inclination allows for controlled filler application, ensuring that only the tip of the rod comes into contact with the weld pool. Additionally, positioning the filler rod diagonally rather than parallel to the joint reduces the risk of heat rising towards the hand holding the rod.
- **Free and steady hand:** Keep your hand holding the filler rod as steady as possible. If necessary, rest the back of your hand on the workpiece or welding table to prevent unwanted movement.

5.   Coordination with the TIG torch

- **Synchronisation:** Coordinating the filler rod movement with the torch movement is essential. Practise coordinated movements where both advance smoothly at the correct pace to maintain a uniform bead.
- **Keep the rod protected:** Whenever you are not adding filler, make sure to keep the rod within the shielding gas area. This will prevent oxidation and ensure it is ready for the next addition.

6.   Techniques for feeding the filler rod

The next section covers some techniques for adding filler metal to the weld pool that will help you improve your control.

"LIGHT TOUCH" TECHNIQUE

In this technique, the objective is to gently touch the edge of the weld pool with the filler rod, rather than introducing it directly into the centre.

- **Starting position:** Hold the filler rod at 15° to 30° relative to the workpiece, depending on its thickness.
- **Touching:** With a controlled movement, bring the tip of the filler rod towards the edge of the weld pool, just touching it lightly.
- **Fusion:** Allow only a small amount of filler metal to melt into the pool.
- **Retraction:** Quickly pull the rod away to prevent it from sticking to the pool.

"CONTROLLED SLIDING" TECHNIQUE

In this technique, the filler rod remains in constant contact with the workpiece, and is slid into the weld pool as the welding progresses.

- **Starting position:** Place the filler rod at a low angle (10° to 15°) relative to the workpiece, keeping it in contact with the material.
- **Sliding:** While advancing with the torch, gradually slide the filler rod into the weld pool in a continuous and controlled manner.
- **Filler application:** As the weld pool moves, the filler rod slides into it, ensuring continuous filler metal addition.
- **Retraction:** If the rod needs to be removed, do so with a quick and clean movement to avoid unwanted adhesion.

"DO NOT FEAR MOVING SLOWLY, FEAR STANDING STILL"

(Chinese proverb)

Every attempt you make, no matter how small, is a step forward. Remember that the value of your effort lies not only in the final result, but also in the dedication and attention you apply to the process. Keep moving forward because with each practice session, you are achieving real success: learning and improving.

Want more information? Watch this exercise in action here:

Facultad de Soldadura: www.youtube.com/watch?v=8yezhSzaLgs

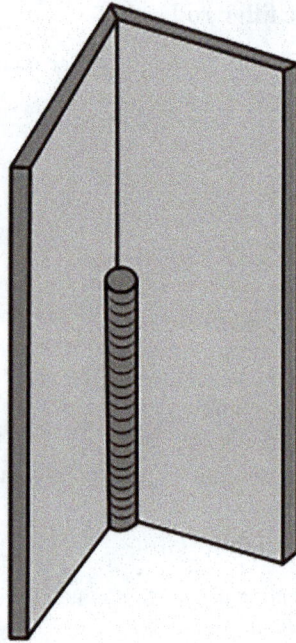

FIGURE 2.10   TIG welding in a vertical-up corner joint. The weld bead is progressively deposited from the base to the top.

TIG WELDING IN CARBON STEEL

PRACTICE 3. VERTICAL-UP FILLET WELD. PF(3F)

- **Base material:** Carbon steel plate 6″ x 1.2″ x ⁵/₁₆ or ⅛″.
- **Electrodes to use:** Choose from available options: Ø 3/32″ thoriated tungsten (WTh20 red), lanthanated (WLa 10/15 black or gold), ceriated (WCe 20 grey), or rare earth (E3 purple).
- **Nozzle:** No.5 or No.6 (⁵/₁₆″ and ³/₈″ diameter, respectively).
- **Filler rod:** Ø 1/16″ or 5/64″, type ER 70S-6.
- **Number of beads:** Two beads, introducing the Walking the Cup technique.
- **Current:** 120–150 amps.
- **Gas flow rate:** 8–10 litres per minute (1 litre per millimetre of nozzle inner diameter).
- **Tungsten stick-out:** ¹/₈″—⁵/₃₂″.

In this practice, we will tackle TIG welding in a fillet joint, now in the vertical-up position, which adds an extra challenge due to the influence of gravity on the welding process (Figure 2.10).

1.   Initial preparation

- **Coupon positioning:** Ensure the coupon is placed at a height where the highest point of the joint remains below your head. This will allow you to maintain a clear view of the weld pool and the tungsten throughout the process, avoiding uncomfortable angles or postures that could hinder visual control.

- **Securing the coupon:** Tack the plates at 90° and firmly secure them to the worktable using, for example, a clamp. This is essential—especially in the vertical-up position—to maintain stability throughout the weld.
- **Cleaning and preparation:** Ensure your tungsten electrode is clean and properly sharpened. Also, wipe the filler rod with a cloth to remove any dirt or contaminants.

2. Torch positioning

- **Torch angle:** Position the torch at 45° relative to both plates, with a slight backward tilt to ensure a clear view of the electrode and weld pool. Keep this angle constant throughout the weld to prevent undercut and ensure a uniform bead.
- **Avoiding handle collisions:** Position the torch so that the handle does not hit the coupon during the Walking the Cup technique. A practical solution is to rotate the torch so that the handle faces upwards, allowing for a smoother movement with no risk of collision.

3. Starting the weld bead

- **Bead initiation:** Start the weld at the bottom of the joint (the base of the vertical joint) and work upwards. Keep the tungsten electrode in a position that allows for full control of the weld pool.

4. Walking the Cup technique

- **Torch position and pressure:** Hold the torch firmly, keeping the torch body at 45°to both plates. The ceramic nozzle must remain in contact with both surfaces of the fillet joint, providing stability and control.
- **Side-to-side movement of the handle:** Move the torch handle side to side, from left to right and right to left. This movement, combined with slight forward pressure, allows the nozzle to progress along the joint without the tungsten getting too close or too far from the workpiece. The progression should be natural and controlled, resulting from both oscillation and pressure.
- **Filler rod angle:** When feeding the filler rod, position it at a slight angle to the joint rather than parallel. This prevents rising heat from bothering your hand and ensures that only the rod's tip melts, allowing better control of the amount of material added to the weld pool.

5. Dry run practice

- Before striking an arc, practise the movement without turning the machine on. Move the torch handle side to side and simulate the progression of the ceramic nozzle along the joint. This will help you gain confidence and improve control before working with the arc and molten pool.

6. Welding exercise (without filler metal)

- Once you feel comfortable with the motion, proceed with the weld, but without adding filler metal. Focus on keeping a consistent weld pool size and on executing the Walking the Cup technique smoothly and evenly.

7.   **Welding with filler metal**

- Repeat the exercise, but this time add the filler rod. Apply what you learned in the first practice, ensuring proper coordination between the torch and the filler metal. Always add the filler at a diagonal angle, positioning it at the edge of the weld pool, not in the centre, to avoid unnecessary cooling.

8.   **Performing the hot pass**

- Once the first bead is complete, apply a second bead—or hot pass, as North American welders call it.
- This second bead reinforces the weld, ensuring greater penetration and a stronger joint. Maintain the Walking the Cup technique and control the filler rod feed, making sure that the second bead fully covers the first one evenly (see Figure 2.11).

9.   **Practising on thinner plates**

- Finally, repeat the weld on a coupon of the same dimensions, but with a thickness of $\frac{1}{8}''$ (3mm) or even $\frac{1}{16}$ (1.5mm).
- This will help you adjust the technique to different thicknesses and improve your control over the torch and filler rod.

FINAL THOUGHTS

Mastering the Walking the Cup technique in the vertical-up position requires practice and patience. The key is to maintain a steady, controlled movement, ensuring that the weld pool remains consistent throughout the joint. Applying the hot pass will reinforce your weld bead and ensure a high-quality joint.

Keep practising and refining your technique for optimal results!

TIG WELDING IN CARBON STEEL

PRACTICE 4. OVERHEAD FILLET WELD (PD/4F)

- **Base material:** Carbon steel plate, 6″ x 1.2″ x $\frac{5}{16}$″ (150 x 30 x 8 mm) or $\frac{1}{8}$″ (3mm).
- **Electrodes:** Choose from available options: Ø 3/32″ (2.4mm) tungsten (WTh20 red for thoriated, WLa10/15 black or gold for lanthanated, WCe20 grey for ceriated, or E3 purple for rare earth).
- **Nozzle:** No.5 or No.6 ($\frac{5}{16}$″ or $\frac{3}{8}$″ (8mm or 10mm) diameter, respectively).
- **Filler rod:** Ø 1/16″ or 5/64″ (1.6mm or 2mm), ER 70S-6.
- **Number of beads:** Two beads, applying the Walking the Cup technique.
- **Current:** 120–150 amps.
- **Gas flow rate:** 17–21cubic feet per hour (8–10L/min) (1L/min per mm of nozzle bore diameter).
- **Tungsten stick-out:** $\frac{1}{8}$″ to $\frac{5}{32}$″ (3–4mm).

Also known as welding in the overhead position, this process presents a unique set of challenges that require precise technique and strict control of the equipment. This position is frequently used in scenarios where workpieces cannot be rotated or repositioned for welding in a more comfortable orientation (Figure 2.12).

FIGURE 2.11 Test piece with a vertical-up corner joint welded using the TIG process. The uniformity of the weld bead and fusion between materials is visible.

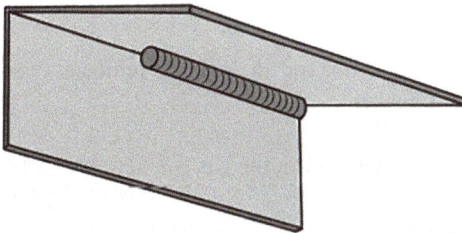

FIGURE 2.12 TIG welding in an overhead fillet joint. The weld bead is visible along the joint between both pieces.

1. Welder positioning

In the overhead position, the welder is positioned directly underneath the joint being welded. Ensure that the highest point of the joint is at a height where you can clearly see the weld pool and the electrode without excessively tilting your head back. Ideally, the joint should be positioned just above eye level, allowing for a comfortable and stable posture.

2. Controlling the electrode and torch

Due to the orientation of the workpiece, gravity affects both the filler metal and molten pool differently. Precise control of the torch is crucial to prevent the tungsten from getting too close to the joint or the weld pool from shifting uncontrollably.

- **Torch angle:** Hold the torch at 45° to the joint, ensuring that the inclination allows a clear view of both the tungsten and the weld pool.
- **Arc length:** In the overhead position, there is an increased risk of molten material sagging. To counteract this, maintain a shorter arc length (approximately ⅛″ (3mm)). This helps focus the heat on the joint, preventing excessive molten metal movement or dripping.

3. Movement and the Walking the Cup technique

Although you have already practised the Walking the Cup technique in other positions, in this overhead position, precise execution is even more critical.

- **Pressure and contact:** Due to gravity, you may need to apply slightly more pressure against the workpiece to keep the torch stable—but avoid pressing too hard. The ceramic nozzle must remain in continuous contact with the plate to provide a steady support.
- **Controlled movement:** Instead of using a wide movement, apply a controlled, precise motion. This ensures that the molten metal stays in place and that the weld bead remains uniform.

4. Filler rod addition

Adding filler metal in the overhead position requires extra caution. The position of the rod and its coordination with the torch are crucial to prevent the filler from detaching or failing to fuse properly.

- **Filler rod angle:** Keep the filler rod at an angle of 15° to 30°, slightly inclined towards the joint, so only the tip of the rod enters the weld pool.

This minimises the risk of heat affecting the hand holding the filler rod and ensures better control of deposition.

- **Synchronization:** Feed the filler rod in small increments, ensuring the material fully melts before advancing further.

The filler rod must always remain within the gas shield to prevent oxidation.

5. Performing the hot pass

Once you have completed the first bead, apply a second bead, or "hot pass". This reinforces the weld and ensures full and even penetration.

- **Purpose of the hot pass:** The hot pass smooths out any irregularities in the first bead and ensures complete fusion at the root of the weld, improving the strength and durability of the joint.
- **Application:** When applying the hot pass, maintain the same parameters as in the first pass, but ensure the new bead fully covers the previous one, blending smoothly into it.

FINAL THOUGHTS

This exercise is particularly demanding and requires high concentration and control. However, with patience and practice, you will master the overhead welding position.

This will prepare you for any welder certification test that requires overhead welding.

Stay motivated and keep pushing forward!

THE ROLE OF THE WELDING INSPECTOR (CSWIP WI) IN TIG WELDING

Inspection protocol for a fillet weld using TIG

When performing a fillet weld using the TIG process, the **CSWIP WI** follows a protocol similar to that used for SMAW, but with specific considerations due to the nature of the TIG process.

1. Visual inspection

The CSWIP WI begins with a visual inspection of the weld, looking for surface defects such as cracks, porosity, undercutting, or excess weld metal.

Since the TIG process produces high-quality welds with a smooth surface finish, any irregularity, no matter how small, can indicate potential issues.

A TIG weld should be uniform, with a flat or slightly convex profile, and free from discontinuities.

2. Dimensional Measurement

The CSWIP WI measures the length, width, and height of the weld bead to ensure compliance with welding procedure specifications.

In the TIG process, it is particularly important to check for excessive weld reinforcement or undercutting, as these are common defects if arc control and filler metal application are not properly managed.

3. Evaluation Based on ISO 5817

The weld is assessed against ISO 5817 criteria, which classifies welding defects into three quality levels:

- B (High quality)
- C (Medium quality)
- D (Low quality)

For welder certification, at least Level C is usually required.

However, since TIG welding is a highly precise process, quality expectations are often more demanding.

Visual inspection protocol

- **Surface defects:** No visible cracks or slag inclusions are allowed. Porosity must be minimal and evenly dispersed. In TIG welding, even the smallest inclusions or pores are easily detectable due to the clarity of the weld bead.
- **Bead profile:** The weld bead must be uniform, without undercutting or excessive reinforcement. Fusion must be complete, and the molten pool must have adequately spread across the joint, ensuring no unfused areas.

DESTRUCTIVE TESTING PROTOCOL

Fracture test of the test coupon

The CSWIP must confirm that weld penetration has reached the required depth within the joint, which is critical in high-responsibility welds.

To verify this, a fracture test is conducted on the test coupon.

Step 1: Removal of tack welds
- The tack welds are removed using a grinder and cutting disc.

Step 2: Cutting the root of the weld bead
- For TIG welding, due to the small weld volume, grinding the root is not necessary before performing the fracture test.

Step 3: Breaking the test coupon
- The coupon is placed in a press or vice and firmly secured.
- Force is applied to the ends of the joint to close the weld and break it, allowing the WSI to examine the internal penetration.

Step 4: Evaluating penetration
- The WSI inspects the fractured edge to measure penetration depth, ensuring it is sufficient for a strong and durable weld.
- A minimum penetration depth of 0.04" (1mm) is recommended.

MEASURING PENETRATION DEPTH

The CSWIP measures penetration using a vernier caliper whenever possible.

- If penetration is insufficient, the weld must be repeated, with adjustments to welding parameters such as arc intensity and travel speed until adequate penetration is achieved.

FINAL CONSIDERATIONS

- Weld stops and tie-ins:
  - o During a certification test, at least one stop with a tie-in is required for each weld bead.
  - o Tie-ins must not coincide in the same location, as they are critical points where defects, such as lack of fusion or porosity, may occur.

NEED A SIMPLER EXPLANATION? HERE IT IS!

CSWIP INSPECTION FOR A TIG FILLET WELD

When you make a fillet weld using the TIG process, an inspector examines your work to ensure it meets quality standards.

The process is similar to other types of welding inspections, but TIG welding has some unique aspects.

1. First visual check

The first step for the inspector is to closely examine the weld.
They are looking for defects such as:

- ☑ Cracks
- ☑ Tiny holes (porosity)
- ☑ Undercutting
- ☑ Overly high weld beads

Since TIG welding produces very clean and smooth welds, any small defect is highly visible. Your weld should look neat and uninterrupted.

2. Checking measurements

Next, the inspector measures your weld to verify its size. This includes the weld length, width, and height.

It's important that the bead isn't too high or has areas where material is missing (undercut), as these problems can occur in TIG welding if arc control and filler metal application are not precise.

3. Comparing with a standard

The inspector compares your weld with ISO 5817, a standard that rates welds into three quality levels:

- B (Best quality)
- C (Medium quality)
- D (Lower quality)

To pass a certification test, you normally need at least Level C.

Since TIG welding is very precise, higher quality levels are often expected.

BREAKING THE WELD TO CHECK ITS QUALITY

FRACTURE TESTING

To verify weld strength, the inspector may perform a "destructive test"—breaking the weld to examine it.

How it's done:

- Step 1: Removing tack welds using a grinder.
- Step 2: Cutting the weld root. For TIG welds, this isn't necessary as the bead is small enough to fracture easily.
- Step 3: Breaking the test piece
  o The welded coupon is clamped in a vice or press.
  o Force is applied to break the weld, revealing the internal penetration.
- Step 4: Checking penetration depth
  o The inspector measures penetration to ensure it is deep enough for a strong joint.
  o Less than 0.04" (1mm)? You need to adjust your welding technique.

MEASURING PENETRATION DEPTH

The inspector uses a vernier caliper to measure penetration depth.

If the weld does not penetrate deeply enough, you must adjust welding settings like arc intensity and travel speed to achieve proper fusion (Figure 2.13).

FINAL NOTES

- **Weld stops and tie-ins:**
  o During a certification test, you must include at least one stop per weld bead.
  o Stops should not be in the same place, as these areas are prone to defects like porosity or lack of fusion.

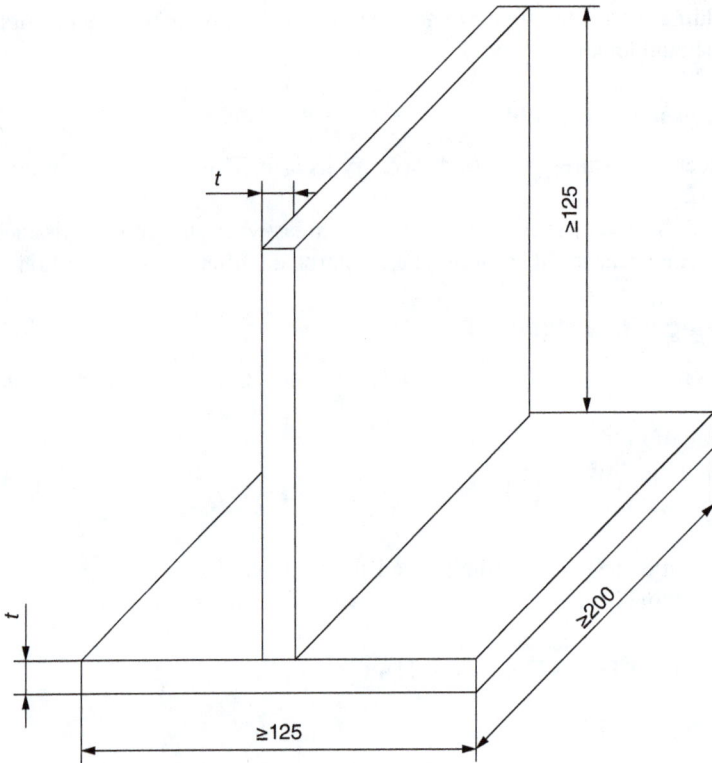

FIGURE 2.13 Dimensional diagram of a fillet joint specimen. The minimum plate dimensions and thickness are specified for a welding test.

Want to practise this exercise on a test coupon with certification dimensions?

Here are the minimum dimensions (in millimetres) specified by UNE EN ISO 9606–1:

TIG WELDING ON CARBON STEEL

PRACTICE 5. V-GROOVE PLATES IN HORIZONTAL POSITION, PA (1G)

Materials and equipment:

- **Base material:** Carbon steel plate, 6″ x 1.38″ x 5/16″ (150 x 35 x 8 mm), bevelled at 35° with no root face, and with a 0.1″ (2.5mm) root gap.
- **Electrodes:** Select from Ø 3/32″ (2.4mm) electrodes, including thorium (WTh20 red), lanthanum (WLa 10/15 black or gold), cerium (WCe 20 grey) or rare earth (E3 purple), depending on availability.
- **Nozzle:** #5 or #6 (Ø 5/16″ and 3/8″ or 8mm and 10mm, respectively).
- **Filler rod:** Ø 3/32″ or 1/8″ (2.4mm or 3.2mm), type ER 70S-6.
- **Number of beads:** Two passes using the Walking the Cup technique (first pass "root" and second pass "hot pass").
- **Current:** 90–120 amps.
- **Shielding gas flow rate:** 17–21 cubic feet per hour (8–10L/min) (1L/min per mm of nozzle inner diameter).
- **Electrode stick-out from nozzle:** 1/8–5/32″ (3–4mm).

1. Initial preparation

Preparing the workpiece:

- **Bevel preparation:** Ensure both plates are bevelled at 35° with no root face (Figure 2.14). This geometry is essential to ensure proper weld penetration, particularly in the root pass. The 0.1″ (2.5mm) root gap is critical to allow correct application of the Lay Wire technique on the first pass.
- **Cleaning:** Before welding, thoroughly clean the plates to remove any rust, grease, or impurities. Use a grinder with a sanding disc or wire brush to ensure the surface is completely free of contaminants. A clean surface is key to preventing weld defects, such as porosity or inclusions.
- **Clamping:** Secure the plates using tack welds at the ends and in the centre if necessary. This will keep the plates in place during welding and prevent warping due to the heat input.

2. Positioning and safety

Workpiece positioning:

- Ensure the test coupon is properly fixed to the worktable, with the highest point of the joint below your eye level. This will allow a clear view of the weld pool while keeping the tungsten electrode in sight, essential for controlling bead quality.
- The **back of the joint must be left open**, without resting on the table or any **support.**

PPE and safety measures:

- Always wear the correct PPE: A welding helmet with an appropriate filter, welding gloves, an apron, and protective sleeves.
- Ensure proper ventilation in the work area to prevent the accumulation of shielding gases.

3. Walking the Cup technique for butt welding

Applying the root pass:

- **Torch positioning:** Hold the torch at 70°–80° relative to the joint, with the tungsten slightly tilted forward. The ceramic nozzle must remain in contact with both the upper and lower edges of the bevel.
- **Walking the Cup movement:** Use the Walking the Cup technique, maintaining a steady and rhythmic lateral movement. The torch must advance along the bevel while applying consistent pressure against the joint walls. This lateral motion helps control the nozzle movement and keeps the tungsten electrode correctly oriented towards the weld pool.

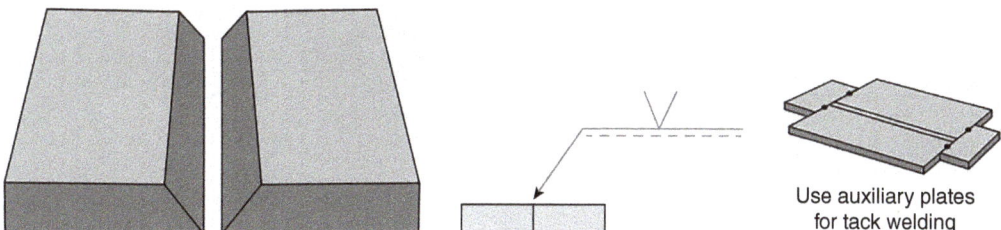

Use auxiliary plates
for tack welding

FIGURE 2.14   Preparation of a V-groove butt joint. Two metal pieces with symmetrical bevels are shown, ready to be welded together.

- **Lay Wire technique:** For the root pass, use the Lay Wire technique to apply filler metal. Position the filler rod over the root gap, ensuring that it is equal to or slightly larger than the gap so that it rests above the opening. As the weld pool advances, it will gradually melt the filler rod, ensuring uniform penetration and preventing cavity formation.

The root of the weld should appear completely clean on the backside, with no concave areas or lack of material, and with a slight reinforcement to ensure full fusion.

Applying the hot pass:

- **Purpose of the hot pass:** This second pass aims to reinforce the root and ensure complete penetration without defects. The term hot pass, commonly used in North American welding jargon, refers to the second pass that seals and reinforces the root bead.
- **Walking the Cup movement**: Just like in the root pass, apply the Walking the Cup technique to maintain even progression and avoid defects like undercutting or lack of fusion.
- **Filler metal application:** Carefully coordinate the filler rod movement with the torch. Always apply the filler rod at the edge of the weld pool to prevent excessive deposition, which could cool the pool too much and compromise weld quality.

4. Completion and inspection

- **Controlled cooling:** Allow the workpiece to cool gradually to prevent residual stresses, which could cause cracks in the weld.
- **Visual inspection:** Check the weld bead to ensure there are no porosity, cracks, or other defects. A good weld should be uniform, free of undercutting, and with complete penetration.

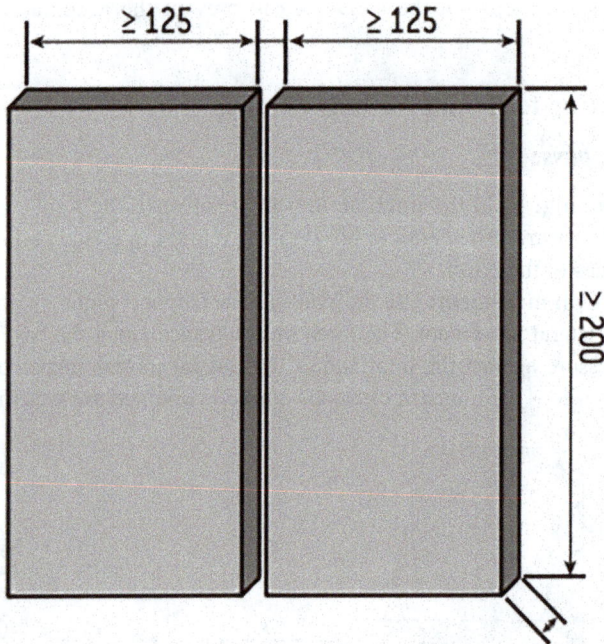

FIGURE 2.15 Dimensional diagram of a butt joint specimen. The minimum plate dimensions and the bevel at the base for welding are indicated.

Final Tips:

☑ **Practice and patience:** Mastering both Walking the Cup and Lay Wire techniques takes continuous practice. Don't get frustrated if you don't achieve perfect results immediately. With time and training, your skill will improve. Practise in dry runs until you master the movement.

☑ **Attention to detail:** Always maintain a comfortable posture and stay fully focused. TIG welding requires precision and constant attention.

Want to practice this exercise on a certification-size test coupon?

Here are the **minimum dimensions (in millimetres)** specified by **UNE EN ISO 9606–1**:

Remember: You must believe you can do it. Do you remember how clumsy I am? If I managed to do it, so can you. Every attempt is a reflection of your best effort (Figure 2.15).

"Dripping water hollows out stone, not through force, but through persistence."

(Latin proverb)

Want more details? Watch the execution of this exercise here:
Facultad de Soldadura: www.youtube.com/watch?v=CZMyNk6THPk

TIG WELDING ON CARBON STEEL

PRACTICE 6: BUTT WELDING OF TWO PLATES IN OVERHEAD POSITION. PC (2G)

Materials and Equipment:

- **Base material:** Carbon steel plate, 6″ x 1.38″ x 5/16″ (150 x 35 x 8 mm), bevelled at 35° with no root face, and with a 0.1″ (2.5mm) root gap.
- **Electrodes:** Select from Ø 3/32″ (2.4mm) electrodes, including thorium (WTh20 red), lanthanum (WLa 10/15 black or gold), cerium (WCe 20 grey) or rare earth (E3 purple), depending on availability.
- **Nozzle:** #5 or #6 (Ø 5/16″ and 3/8″ or 8mm and 10mm, respectively).
- **Filler rod:** Ø 3/32″ or 1/8″ (2.4mm or 3.2mm), type ER 70S-6.
- **Number of beads:** Two passes using the Walking the Cup technique (first pass "root" and second pass "hot pass").
- **Current:** 90–120 amps.
- **Shielding gas flow rate:** 17–21 cubic feet per hour (8–10L/min) (1L/min per mm of nozzle inner diameter).
- **Electrode stick-out from nozzle:** 1/8–5/32″ (3–4mm).

Overhead position welding (PC) presents a greater challenge than horizontal welding due to gravity, which causes molten metal to flow downward (Figure 2.16). For this reason, it is essential

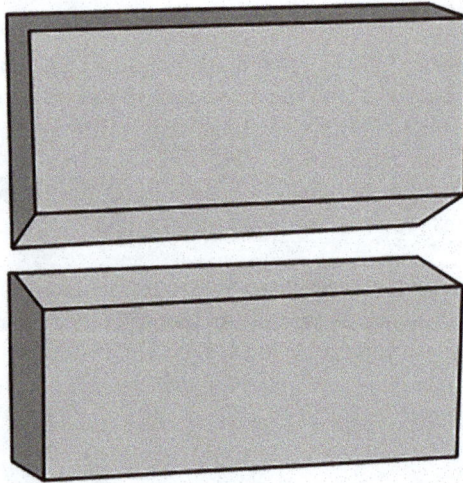

FIGURE 2.16   Preparation of a V-groove butt joint. Two metal pieces with bevels are shown, ready to be welded together.

to adjust welding parameters correctly and practise Walking the Cup and Lay Wire techniques in dry runs before proceeding with actual welding.

In this position, the root bead must be perfectly clean on the backside of the joint, with no concave areas or lack of fusion, and with a slight reinforcement ensuring full and uniform fusion.

1.   Preparing the workpiece

Before starting, ensure the plates are properly prepared and securely fixed in position. Check that the 35° bevel is clean and free of oxides or contaminants.

The 0.1″ (2.5mm) root gap is crucial to allow for proper penetration of the root bead, especially when using the Lay Wire technique, which involves keeping the filler rod in constant contact with the bevel while executing the weld.

2.   Walking the Cup technique in overhead position

In this position, the Walking the Cup technique must be adapted as follows:

- **Dry run practice:** Before starting the weld, it is highly recommended to practise the movement without an arc. This will help you get used to the motion in this specific position, where gravity affects the weld pool and filler rod differently.
- **Torch positioning:** Keep the torch at 45° relative to the plates, ensuring that the ceramic nozzle maintains constant contact with both sides of the bevel. This contact is essential to control the weld progression and ensure adequate gas shielding.
- **Lateral movement:** Move the torch laterally with gentle, controlled motions, avoiding excessive downward pressure that could bring the tungsten electrode too close to the weld pool. The torch should advance slowly and steadily, ensuring the weld pool remains even throughout the joint.
- **Coordination with the filler rod:** When using the Lay Wire technique, the filler rod must remain in continuous contact with the bevel while the torch advances. This helps maintain consistent filler metal deposition without needing to lift the rod from the pool. The key is to synchronise the torch movement with the filler rod advance to avoid excess material accumulation or lack of fusion.

3. Controlling the root bead

The main goal in this practice is to achieve a clean, well-formed root bead on the backside of the joint. The root bead should exhibit full fusion, with no concave areas or lack of material.

A slight reinforcement is acceptable, provided that the surface remains uniform and defect-free. Pay close attention to:

- Welding speed to prevent overheating.
- Arc length to avoid defects.
- Gravity's effect on molten metal flow to prevent bead sagging.

4. Executing the hot pass

After completing the root bead, proceed with a second bead known as the hot pass, the weld pass deposited over the root bead to reinforce it and ensure a strong joint.

- Hot pass execution: The hot pass should be carried out using the Walking the Cup technique just as in the root bead, but with a slightly higher amperage.
- Objective: Ensure proper fusion between the root bead and the hot pass, eliminating any imperfections that may have formed during the first pass.

5. Additional exercise with 1/8″ (3mm) plates

Once you have successfully completed the weld on 5/16″ (8mm) plates, it is recommended to repeat the same exercise using 1/8″ (3mm) thick plates, bevelled at 35° with no root face.

Welding thinner materials will require adjustments to welding parameters, such as:

- Lower amperage.
- Greater control of technique to avoid burn-through or excessive distortion.

This exercise will help refine your skill in TIG welding in the overhead position, preparing you for more demanding situations on thinner materials.

TIG WELDING ON CARBON STEEL

PRACTICE 7. BUTT WELDING OF TWO PLATES IN VERTICAL UPWARD POSITION. PF (3G)

Materials and equipment:

- **Base material:** Carbon steel plate, 6″ x 1.38″ x 5/16″ (150 x 35 x 8 mm), bevelled at 35° with no root face, and with a 0.1″ (2.5mm) root gap.
- **Electrodes:** Select from Ø 3/32″ (2.4mm) electrodes, including thorium (WTh20 red), lanthanum (WLa 10/15 black or gold), cerium (WCe 20 grey) or rare earth (E3 purple), depending on availability.
- **Nozzle:** #5 or #6 (Ø 5/16″ and 3/8″ or 8mm and 10mm, respectively).
- **Filler rod:** Ø 3/32″ or 1/8″ (2.4mm or 3.2mm), type ER 70S-6.
- **Number of beads:** Two passes using the Walking the Cup technique (first pass "root" and second pass "hot pass").
- **Current:** 90–120 amps.

- **Shielding gas flow rate:** 17–21 cubic feet per hour (8–10L/min) (1L/min per mm of nozzle inner diameter).
- **Electrode stick-out from nozzle:** 1/8–5/32" (3–4mm).

Vertical upward welding (PF) presents an additional challenge due to gravity acting against the upward movement. This requires strict control of the weld pool and travel speed to avoid issues such as molten metal sagging (Figure 2.17).

Before starting the weld, it is highly recommended to practise dry runs of the Walking the Cup and Lay Wire techniques in this position to adapt to the specific gravitational challenges.

2. Walking the Cup technique in vertical upward position

The Walking the Cup technique in this position requires greater control to counteract gravity's effect:

- **Controlled upward movement:** Unlike horizontal welding, movement must be slower and more controlled. Ensure the weld pool stays in place and does not start flowing downward. Keep the torch at a slightly higher angle to direct heat towards the upper part of the pool.

If the pool starts to sag, reduce the torch angle to prevent excessive filler metal from melting, which could lead to over-thick weld beads and tungsten contamination.

- **Steady and slightly slower travel speed:** The torch advance should be methodical to ensure uniform metal deposition. The filler rod should remain in continuous contact with the weld pool, moving at the correct pace to achieve full fusion while avoiding excessive material buildup.

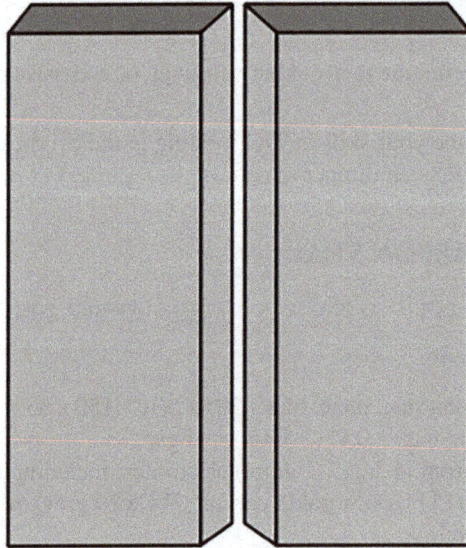

FIGURE 2.17 Preparation of a V-groove butt joint. Two metal pieces with bevels are shown, ready to be welded together.

- **Coordination with the filler rod:** In the Lay Wire technique, the filler rod must stay in constant contact with the weld pool and advance at a steady pace to ensure proper fusion. To reduce excess filler metal, introduce the rod diagonally rather than parallel to the joint.

3. Controlling the root bead and hot pass

In this position, the backside of the root bead should be just as clean as in previous practices, with a slight reinforcement and no concave areas.

The hot pass reinforces the root bead and must be performed with equal precision, ensuring that both beads integrate perfectly.

Using the Walking the Cup technique in the hot pass will help:

- Eliminate any remaining imperfections from the root pass.
- Ensure a uniform weld bead.

4. Additional exercise with 1/8″ (3mm) plates

For an extra challenge, repeat this exercise using 1/8″ (3mm) thick plates, bevelled at 35° with no root face.

Welding thinner materials requires:

- More precise control of the weld pool.
- Adjusted travel speed to prevent burn-through.
- Careful application of the Lay Wire technique to avoid excessive material buildup.

This exercise will help refine your technique for more demanding applications in vertical TIG welding.

INSPECTION PROTOCOL FOR A CSWIP WELDING INSPECTOR (CSWIP WI) IN A V-GROOVE BUTT WELD ON 5/16″ (8MM) CARBON STEEL USING TIG

1. Visual inspection

The initial visual inspection is similar to that of fillet welds, but for butt joints, special attention must be paid to the following aspects:

- **Weld bead profile:** In a V-groove butt weld, the bead must completely fill the joint without leaving cracks at the root. The weld should be fully fused and uniform along the bevel, with no gaps or lack of penetration at the root. The bead should be even and slightly convex.
- **Penetration:** Penetration is critical in this type of joint to ensure structural strength. The CSWIP WI will check for signs of insufficient fusion or lack of penetration, which may appear as an irregular bead profile.
- **Fusion of the bevel edges:** The weld must extend across the entire bevelled surface, ensuring no unfused areas, particularly at the root. Any lack of fusion at this stage would compromise the integrity of the joint.

2. Measurement of dimensions and geometry

For V-groove butt welds, dimensional measurements are key to ensuring that penetration is sufficient and that the weld bead meets Welding Procedure Specification (WPS) requirements:

- **Weld bead dimensions:** The CSWIP WI will measure the bead height, width, and length to confirm they are within the WPS parameters. Excess reinforcement in a butt joint can introduce additional stress, so strict control is necessary.
- **Root bead penetration measurement:** The minimum penetration must be within the specified range (typically 0.02″ to 0.08″ (0.5mm to 2mm)). Insufficient penetration can weaken the joint's strength.

3. Evaluation according to ISO 5817 standard

As with fillet welds, the butt weld will be assessed based on ISO 5817 criteria, with a minimum quality level of C required. However, for butt joints, defects such as lack of penetration and undercut along the bevel edges are more critical since they can directly affect the structural integrity of the joint.

NON-DESTRUCTIVE AND DESTRUCTIVE TESTING PROTOCOL

1. Radiographic testing

For a **V-groove butt joint TIG-welded on carbon steel**, the **radiographic inspection** follows these steps:

1. **Preparation:** The test coupon is cleaned to avoid any interference in the image and positioned correctly alongside a radiographic film or digital detector.
2. **Exposure to radiation:** An X-ray or gamma-ray source is used to pass radiation through the weld, capturing an image on the film or detector.
3. **Image capture:** The film is chemically developed, or the digital detector records the radiographic image of the weld.
4. **Inspection of the radiograph:** The image is analysed to detect internal defects such as porosity, cracks, inclusions, or lack of fusion, following applicable inspection standards.

Radiographic testing provides a highly precise non-destructive method for verifying internal weld quality.

2. Fracture testing of the test coupon

To verify penetration and internal fusion, the CSWIP WI may conduct a destructive test on the sample coupon. However, for butt joints, fracture testing is often replaced by macrographic and micrographic examinations.

MACROGRAPHIC EXAMINATION IN A V-GROOVE BUTT JOINT WELDED WITH TIG

1. **Sample preparation:** A cross-section of the welded coupon is cut to expose the welded joint.
2. **Surface polishing:** The cut surface is polished until smooth and free of scratches, ensuring clear detail visibility.

3. **Chemical etching:** A chemical reagent (e.g., nitric acid diluted in alcohol) is applied to reveal the weld structure and characteristics.
4. **Visual inspection:** The joint is examined with the naked eye or a magnifier to verify:
   o Proper fusion of the root and bevel edges.
   o Uniformity of weld beads.
   o Absence of internal defects such as porosity, cracks, or inclusions.

Macrographic analysis is essential to determine if the welder and the welding procedure meet the required quality standards. This test is often combined with bending tests.

MICROGRAPHIC EXAMINATION IN A V-GROOVE BUTT JOINT

1. **Cutting and sample extraction:** A representative section of the welded coupon is cut, ensuring it includes the base metal, heat-affected zone (HAZ), and weld metal.
2. **Polishing:** The sample is gradually polished using finer abrasive papers (down to 1200 grit or higher), followed by fine abrasive paste polishing to achieve a completely smooth, scratch-free surface.
3. **Chemical etching:** A suitable reagent (e.g., 2–5% Nital for carbon steel) is applied to highlight microstructures, distinguishing the base metal, HAZ, and weld metal.
4. **Microscopic examination:** The sample is viewed under a metallurgical microscope (50x–500x magnification) to identify:
   o Grain size.
   o Defects (cracks, porosity).
   o Microstructure distribution (ferrite, pearlite, bainite).
5. **Evaluation:** Observations are documented and checked against applicable standards (e.g., AWS D1.1, ISO 9606) to verify compliance.

3.   Bend test

For a V-groove butt joint welded with TIG, the bend test follows these steps:

1. **Preparation of the test specimen:** A section of the welded coupon is cut, typically rectangular, including the weld and base material.
2. **Placement in the bending machine:** The specimen is positioned in a bend testing machine with the appropriate rollers or mandrel.
3. **Bending application:** The specimen is bent to a specified angle (usually 180°), with the weld bead either on the tensioned (face bend) or compressed (root bend) side.
4. **Evaluation:** The bent specimen is inspected for defects such as cracks or fractures, which indicate weld failure. Standards define the acceptance criteria, including maximum allowable defect length or depth.

This test evaluates the ductility and overall quality of the weld under deformation.

FINAL CONSIDERATIONS

- **Weld stops and tie-ins:** In butt welds, it is essential that tie-ins are properly executed, particularly at the root. Any defect in the tie-ins can cause lack of fusion, which is more critical in butt joints than in fillet welds.
- **Weld continuity evaluation:** The CSWIP WI must verify that there are no major discontinuities in the bead. The joint must be uniform along its entire length, with no sudden changes in bead height or width that could compromise strength.

This inspection protocol focuses on the specific requirements for 5/16" (8mm) bevelled butt joints, particularly the importance of complete penetration and fusion at the root.

While it shares some steps with fillet weld inspections, the critical nature of penetration and bead continuity assessment requires a stricter inspection process for this type of joint.

Oops . . . I got a bit too technical again! Let me give it another shot in plain English:

Visual inspection:

1. **Check the weld bead:** The weld bead must completely fill the joint, leaving no gaps or cracks at the root (the deepest part of the joint). It must be uniform, with a slight convex shape.
2. **Penetration:** It is crucial that the weld fully penetrates the root to ensure strength. If not, the weld may appear irregular or have unjoined areas.
3. **Fusion:** The entire bevelled surface must be fully fused, with no unfused areas. This ensures a solid joint.

Measurement of dimensions and shape:

1. **Weld bead size:** The height, width, and length of the bead are measured to ensure they meet the procedure's requirements. A bead that is too large or too small can cause problems.
2. **Root penetration:** The depth of penetration is checked (typically between 0.02" and 0.08" (0.5mm to 2mm)). Insufficient penetration can weaken the weld.

Evaluation according to standards:

- The weld must comply with the quality standards set by ISO 5817.
- Common problems to avoid include lack of penetration or poorly fused edges, as they affect the weld's strength.

Non-destructive and destructive testing:

Radiography:

- A radiographic test (X-ray or gamma-ray) is used to detect internal defects such as voids, cracks, or lack of fusion, without damaging the weld.

Fracture or macrographic examination:

Macrography:

1. A section of the welded material is cut.
2. The cut surface is polished until smooth, then a chemical reagent is applied to reveal details.
3. The weld is examined to check root and edge fusion, ensuring there are no internal defects.

Micrographic Examination:

- Similar to macrography, but using a microscope to observe finer details, such as:
  o Grain size
  o Internal structure
  o Small defects like tiny cracks or inclusions
- This ensures the weld is fully fused at a microscopic level.

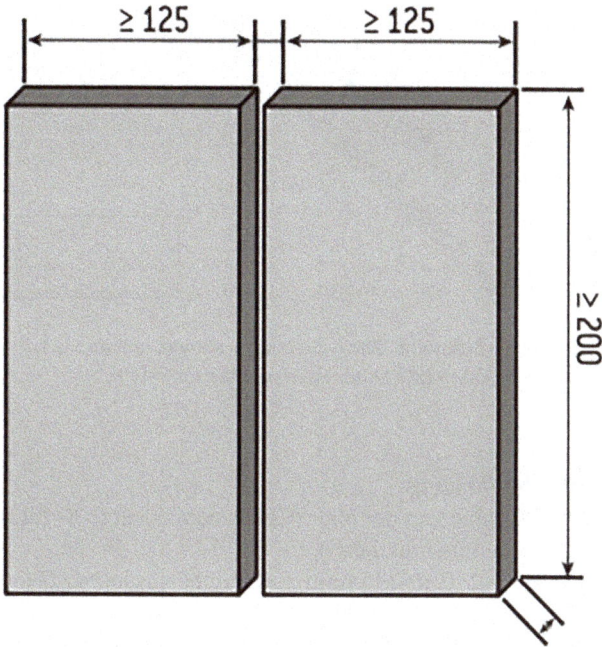

FIGURE 2.18 Dimensional diagram of a butt joint specimen. The minimum plate dimensions and the bevel at the base for welding are indicated.

Bend Test:

1. A sample of the weld is cut and bent into an arc or even a tight angle.
2. If cracks or fractures appear, it means the weld is not strong enough.

Final Considerations:

- Good tie-ins between weld beads are essential, especially at the root. Any defects here could weaken the joint.
- The weld must be uniform, without sudden changes in shape that could create weak points.

If you want to practise this exercise on a qualification test coupon, the minimum dimensions (in millimetres) according to UNE EN ISO 9606–1 are as follows (Figure 2.18):

TIG WELDING ON CARBON STEEL

PRACTICE 8. BUTT WELDING SHEETS—HORIZONTAL POSITION PA (1G)

- **Base material:** Carbon steel plate, 3.94″ × 1.38″ × 0.06″ (100 × 35 × 1.5 mm).
- **Electrodes:** Choose from Ø 3/32″ (2.4mm) tungsten electrodes, either thorium (WTh20 red), lanthanum (WLa 10/15 black or gold), cerium (WCe20 grey), or rare earths (E3 purple).
- **Gas nozzle:** No. 5 or No. 6 (Ø 5/16″ and Ø 3/8″, respectively).
- **Filler rod:** Ø 1/16″ or Ø 5/64″ (1.6mm or 2mm), ER 70S-6 type.
- **Number of weld beads:** One bead.

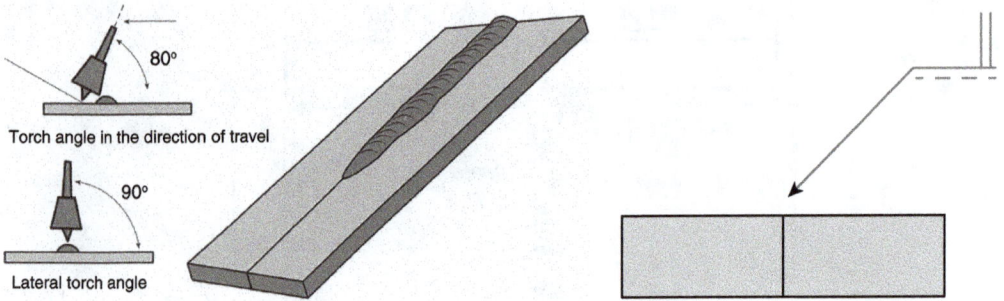

FIGURE 2.19   TIG welding in a butt joint. The weld bead is shown, and on the left, diagrams indicate the torch inclination in the travel direction (80°) and lateral inclination (90°).

- **Current intensity:** 50–70 amps.
- **Gas flow rate:** 16–20 cubic feet per hour (CFH) (equivalent to 8–10L/m, following the rule of 1L/mm of nozzle inner diameter).
- **Tungsten extension:** 0.12–0.16″ (3–4mm) beyond the gas nozzle (Figure 2.19).

Step-by-step guide: TIG welding of 1.5mm sheets in horizontal position

1. **Preparing the work area**

   - Ensure your workspace is clean and organised, with enough room for manoeuvring during welding. It is crucial that no contaminants affect the weld quality.

2. **Cleaning the surfaces to be welded**

   - Thoroughly clean the plates with a clean cloth to remove dust and dirt. If necessary, use a degreaser to eliminate any oil or grease residue.
   - Sand the edges of the sheets (approximately 0.4″/1cm) to remove any oxide or impurities, ensuring a high-quality weld.

3. **Selecting and preparing the electrode**

   - Choose the appropriate electrode, whether thorium, lanthanum, cerium, or rare earths, based on availability.
   - Ensure the electrode is sharpened to a conical point and that the tungsten extends 3–4mm beyond the nozzle. This is essential for maintaining a stable and focused arc.

4. **Positioning the plates and tack welding protocol**

   - Position the plates horizontally, aligning them precisely to ensure a uniform weld.
   - Tack weld the plates at several points along the joint, particularly at the ends and centre. This prevents distortion caused by heat during welding. The tack welds should be small to avoid interfering with the progress of the welding bead.

5. **Adjusting the welding machine**

   - Set the TIG welding machine with the following parameters:
     o Pre-gas: 1-second gas ramp-up to purge the welding area before striking the arc.

o   Initial current ramp: 30% to 100% of the selected current over 3 seconds. This mini-
    mises burn-through at the start.
o   Final current ramp: Gradual decrease from 100% to 30% over 3 seconds. This con-
    trols heat dissipation at the end of the weld.
o   Post-gas: 10-second post-gas ramp to protect the tungsten while cooling, preventing
    oxidation.
- Select the 4T mode on the machine for better process control.

6.   Welding with the torch in the air (without Walking the Cup)

- Position the torch at the start of the joint, approximately 0.4″ (1cm) from the edge.
- Initiate the weld while keeping the tungsten tip 0.08–0.12″ (2–3 mm) from the sheet, at
  an inclination of 10–15 degrees in the direction of travel.
- Progress slowly, maintaining a controlled molten pool. The filler rod can be kept in the
  air, ensuring that only the rod tip enters the molten pool to avoid sudden cooling.

7.   Welding with the rod resting on the joint

- If you prefer resting the filler rod on the joint, keep it at an angle of 15° to 30° degrees
  relative to the workpiece.
- Advance the rod as the molten pool moves forward, ensuring that only the tip of the rod
  touches the pool's edge. This ensures controlled deposition of material without over-
  loading the weld pool.

8.   Executing the weld using Walking the Cup

Striking the arc

- Position the torch 0.08–0.12″ (2–3mm) from the sheet, at 10° to 15° incline in the direction of travel.
- Start the arc as previously explained, ensuring that the initial current ramp is active for
  a smooth start.

Preparation for Walking the Cup

- Ensure the ceramic nozzle is clean and in good condition. Any imperfections may cause
  irregular friction, making movement difficult.
- Place the torch so that the ceramic cup makes contact with the sheet surface. The key
  here is to apply light but constant pressure, avoiding excessive force.

Controlled motion

Tip to prevent slipping or getting stuck: Imagine the gas nozzle as a car on a road. If you push
too hard, it may skid; if you apply too much pressure, it may stop. Find a balance where you feel
stable contact without excessive force.

- Move the torch in a controlled oscillating motion, rocking the nozzle from side to side
  while advancing slowly along the joint.
- Maintain a steady pace, and do not rush. If the nozzle slips, slightly reduce the downward pressure. If the nozzle gets stuck, ease off the pressure or adjust the torch angle.
- Practice this movement before striking the arc to get used to the sensation. This will
  help you develop precise control when welding for real.

Walking the Cup over the bead

- As you advance, the ceramic cup will glide over the weld bead rather than a flat surface.
- **Tip to avoid anxiety:** Imagine the torch gently "stepping" over the bead rather than pushing it forward. If you maintain steady pressure and smooth movement, the nozzle will walk over the bead with ease. If you feel uncertain, slow down and focus on keeping a consistent rhythm.

9. Completing the bead

- As you approach the end of the joint, reduce travel speed slightly while maintaining the same oscillating motion. Activate the final current ramp, pressing the trigger 1cm before reaching the edge.
- Finish the weld with controlled motion, allowing the post-gas ramp to protect the tungsten as it cools.

10. Visual inspection

- After welding, inspect the bead visually to check for defects such as undercut, porosity, or lack of fusion.
- Ensure the root of the bead (on the back side) is clean, with no concave areas and a slight reinforcement for proper fusion.

If you want to practice this exercise on a qualification test coupon, the minimum dimensions (in millimetres) according to UNE EN ISO 9606–1 are as follows (Figure 2.20):

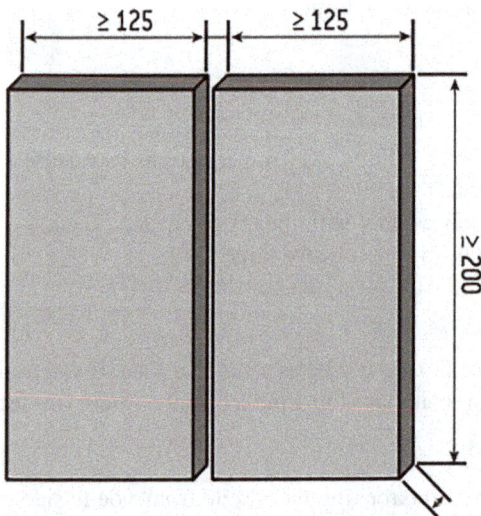

FIGURE 2.20 Test piece with a TIG-welded butt joint. The weld bead is visible with a uniform finish and proper fusion between the pieces.

"The arrow that hits the target is the result of a hundred that missed."

(African proverb)

TIG WELDING ON CARBON STEEL

PRACTICE 9. BUTT WELDING SHEETS—CORNICE POSITION PC (2G)

- **Base material:** Carbon steel plate, 3.94″ × 1.38″ × 0.06″ (100 × 35 × 1.5 mm).
- **Electrodes:** Choose from Ø 3/32″ (2.4mm) tungsten electrodes, either thorium (WTh20 red), lanthanum (WLa 10/15 black or gold), cerium (WCe20 grey), or rare earths (E3 purple).
- **Gas nozzle:** No. 5 or No. 6 (Ø 5/16″ and 3/8″, respectively).
- **Filler rod:** Ø 1/16″ or 5/64″ (1.6mm or 2mm), ER 70S-6 type.
- **Number of weld beads:** One bead.
- **Current intensity:** 50–70 amps.
- **Gas flow rate:** 17–21 cubic feet per hour (16–20 CFH) (equivalent to 8–10L/min, following the rule of 1L per mm of nozzle inner diameter).
- **Tungsten extension:** 1/8–5/32s (3–4mm) beyond the gas nozzle.

TRANSITIONING FROM HORIZONTAL TO CORNICE POSITION WELDING

The most significant change when moving from horizontal welding to the cornice position is how gravity affects the molten pool (Figure 2.21). In this position, the weld tends to flow downward, making it more challenging to control the molten pool and maintain bead uniformity.

EQUIPMENT SETTINGS

CURRENT INTENSITY AND RAMP ADJUSTMENTS

Since the cornice position is more prone to droplet formation due to gravity, it is essential to ensure proper current ramp settings (both at the start and end). This is even more critical than in the horizontal position, as a sudden increase in current could create a molten pool that is too large and difficult to control.

- Keep the ramps smooth, starting with the lower end of the recommended intensity range, and increase only if necessary.

WELDING TECHNIQUE

GAS PURGE AND PRE–/POST–FLOW

- Due to the orientation of the workpiece, ensure that the gas purge is sufficient to protect the weld zone from atmospheric contamination, especially on the lower side of the joint, which may be harder for the shielding gas to reach.
- Set the pre-gas ramp to 1 second to ensure that the area is fully purged before striking the arc.
- Maintain a post-gas flow of 10 seconds to protect the tungsten from oxidation while cooling.

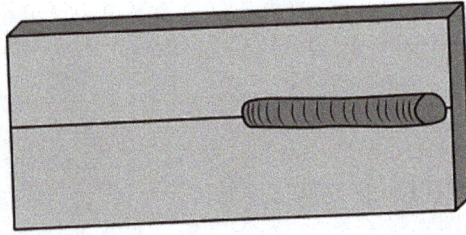

FIGURE 2.21   TIG welding in a butt joint. The weld bead is visible along the joint between the pieces.

MOLTEN POOL CONTROL AND TORCH MOVEMENT

- **Without Walking the Cup:** Move the torch smoothly and steadily, ensuring that the molten pool does not become too large.
- **With Walking the Cup:**
  - o In the cornice position, performing the Walking the Cup technique can be more challenging due to the tendency of the ceramic cup to slip or move unevenly over the smooth surface.
  - o To overcome this, apply slight downward pressure while moving the torch side to side.
  - o This will help stabilise the torch and control the molten pool more effectively.
  - o It is normal to feel that the torch does not move as smoothly as in other positions; however, with practice, you will gain confidence in this movement.

FILLER ROD HANDLING

- **Rod in the air:**
  - o If you choose to keep the filler rod in the air, ensure that you introduce it gently at the edge of the molten pool, avoiding the centre to prevent excessive cooling.
  - o Maintain constant control over the amount of filler added and make small adjustments in the rod's inclination if the molten pool starts to overflow or flow downward.
- **Rod resting on the joint:**
  - o Alternatively, you can rest the rod on the joint and gradually move it towards the edge of the molten pool.
  - o This method can offer greater precision in filler deposition, which is especially useful in the cornice position, where gravity may cause the molten pool to expand more quickly.

FINALISING THE WELD

CURRENT RAMP DOWN

- As you approach the end of the joint, initiate the current ramp-down (3 seconds) at least 10mm before the final edge to prevent burn-through or excessive material build-up.
- Maintain full control of the torch and filler rod during this process to ensure a clean, defect-free finish.

TIG WELDING ON CARBON STEEL

PRACTICE 10. BUTT WELDING SHEETS—VERTICAL-UP POSITION PF (3G)

Base material: Carbon steel plate, 3.94" x 1.38" x 0.06" (100 x 35 x 1.5 mm).

Electrodes to use: Choose according to availability from Ø 3/32" (2.4mm) thorium (WTh20 red), lanthanum (WLa 10/15 black or gold), cerium (WCe 20 grey), or rare earths (E3 purple).

Nozzle: No. 5 or No. 6 (Ø 5/16–3/8"/8–10mm).

Filler rod: Diameter 1/16" or 5/64" (1.6mm or 2mm), ER 70S-6 type.

Number of passes: One pass.

Current: 50–70 amps.

Gas flow rate: 17–21 cubic feet per hour (8–10L/min).

Tungsten stick-out: 1/8–5/32" (3–4mm).

Welding in the vertical-up position presents an additional challenge due to the influence of gravity, which acts against the welding progression. This requires more precise control of the molten pool and filler material technique to avoid defects such as sagging or lack of fusion (Figure 2.22).

EQUIPMENT SETUP

CURRENT INTENSITY AND RAMP SETTINGS

It is essential to optimise the current ramp settings for this type of welding.

- **Start-up:** Use a soft ramp-up (30% to 100% in 3 seconds) to prevent the molten pool from becoming too large and difficult to handle.
- **End of welding:** Apply a similar ramp-down (100% to 30% in 3 seconds) to ensure a clean weld finish.

PRE-GAS AND POST-GAS SETTINGS

- Set the pre-gas ramp to 1 second to ensure the arca is fully purged before starting.
- A post-flow of 10 seconds remains essential to protect the tungsten while cooling, preventing oxidation and ensuring a clean electrode for future work.

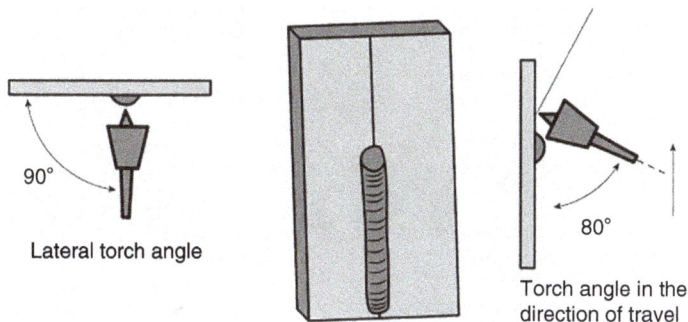

90°

Lateral torch angle

80°

Torch angle in the direction of travel

FIGURE 2.22 TIG welding in a vertical-up butt joint. On the left, a diagram shows the torch's lateral inclination at 90°. On the right, the travel direction inclination is indicated at 80°.

WELDING TECHNIQUE

MOLTEN POOL CONTROL AND TORCH MOVEMENT

1.   Without the Walking the Cup technique

In the vertical-up position, it is crucial to keep the torch angled slightly towards the upper part of the molten pool to maintain visibility of the tungsten.

- The welding progression must be careful, avoiding a too-large molten pool, which could become uncontrollable.
- A useful strategy is to rest one or more fingers of the hand holding the torch on the workpiece for stability and to reduce hand tremor.
  - o To avoid burns, you can use fire-resistant fabric finger sleeves, known as "Fingers TM", available online.
  - o Alternatively, you can make your own sleeve using a welding glove finger, filling it with heat-resistant fabric (see Figure 2.23).

USING THE WALKING THE CUP TECHNIQUE

Applying this technique in this position can be more challenging due to the need for strict control of the molten pool direction.

- **Experiment with different torch positions, including inverting it by positioning the handle upwards.**
  - o This adjustment can create a sensation of added weight on the ceramic nozzle, which may make the Walking the Cup technique feel more natural and easier to control.
- **Maintain a constant and moderate forward pressure, combined with a smooth, controlled side-to-side motion.**
  - o This prevents the molten pool from sliding downwards, ensuring a uniform weld bead.
- **If the molten pool begins to sag, reduce the torch attack angle.**
  - o This will limit the amount of molten metal deposited, preventing excessive bead thickness and tungsten contamination.

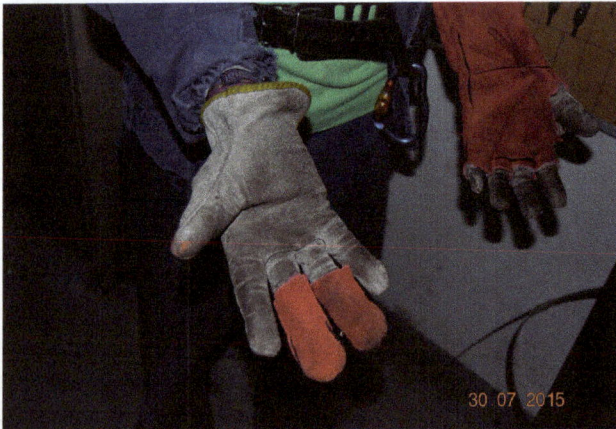

FIGURE 2.23   Short TIG welding gloves partially covered with electric welding gloves, providing extra protection for the fingers and back of the hand.

FILLER ROD CONTROL

- Whether keeping the filler rod airborne or resting it on the joint, always introduce it at the edge of the molten pool, never in the centre, to prevent excessive material deposition and undesired cooling.
- In the vertical-up position, it is recommended to angle the filler rod diagonally to reduce the amount of material deposited and maintain control of the molten pool.
- If you notice the filler rod melting too quickly, forming a large droplet at the tip, or causing excess material accumulation, adjust the rod angle to limit the amount fed into the molten pool.

RAMP-DOWN AND THE "RAT TAIL" EFFECT

As you approach the end of the joint, ensure you initiate the ramp-down precisely when reaching the final edge of the weld.

- Then, move back about 3/8″ (10mm) in the opposite direction as the current decreases.
- This movement creates a progressive narrowing of the weld bead, known as the "rat tail" effect, which resembles the shape of a rodent's tail.

This effect ensures a clean and controlled weld termination, preventing material buildup and perforations at the final edge (Figure 2.24).

TIG WELDING ON CARBON STEEL

PRACTICE 11. PIPE TO PLATE IN HORIZONTAL POSITION—PB (2F)

- **Base material:** Carbon steel plate 4″ x 4″ x 1/8″. Pipe 2″ Ø x 1-3/8″ x 1/8″.
- **Electrodes to use:** Choose from available options: 3/32″ thoriated tungsten (WTh20 red), lanthanated (WLa 10/15 black or gold), ceriated (WCe 20 grey), or rare earth tungsten (E3 purple).
- **Nozzle:** No. 5 or No. 6 (5/16″ and 3/8″ diameter, respectively).
- **Filler rod:** 5/64″ or 3/32″ diameter, type ER 70S-6.
- **Number of beads:** 2 beads (including a hot pass).
- **Current intensity:** 90–120 amps.
- **Gas flow rate:** 17–21 cubic feet per hour (CFH).
- **Tungsten extension beyond the nozzle:** 1/8–5/32″.

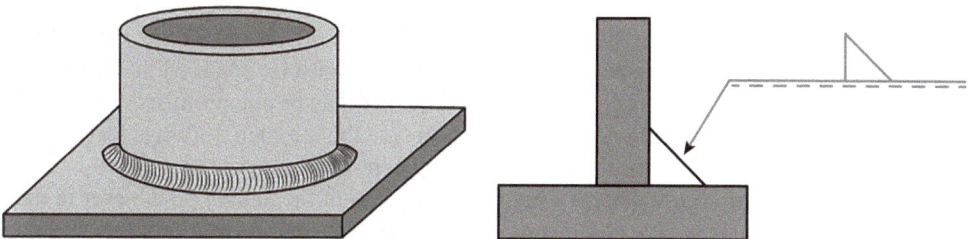

FIGURE 2.24 TIG welding of a tube to a plate in a fillet joint. The weld bead is visible around the base of the tube, securing the connection to the plate.

PREPARATION OF THE WELDING SURFACES

CLEANING AND PREPARATION

Before starting, ensure both the pipe and plate surfaces are completely clean. Remove any rust, paint, oil, or dirt using a wire brush, sandpaper, or a metal degreaser. This guarantees proper weld adhesion and prevents defects during the process. Also, clean the area where the nozzle and filler rod will contact the material—approximately 3/8″ along the entire circumference of the pipe and the surface of the plate.

TACKING THE PIECE

Position the pipe on the plate, ensuring it is perfectly aligned and centred. Apply at least four evenly spaced tack welds around the pipe's circumference to keep it in place. These tacks should be strong enough to prevent movement during welding.

POSITIONING THE COUPON

Once tacked, place the assembly on your workbench so that the joint is in a comfortable position for welding. Remember that ergonomics is key: you should be able to move smoothly around the pipe without straining.

PERFORMING THE WELD

APPLYING THE WALKING THE CUP TECHNIQUE

1. **Head positioning:** Always keep your head between the torch and the filler rod. This allows you to clearly see the tungsten and weld pool. If the torch moves ahead of your head, you risk losing sight of the tungsten, which could result in poor-quality welding.
2. **Rotational movement:** Imagine the pipe as an analogue clock. Start at 12 o'clock and move clockwise until reaching 6 o'clock, passing through 3 o'clock. This rotational movement helps maintain a consistent attack angle, facilitating the Walking the Cup technique. The key is to move smoothly around the pipe while maintaining constant control over the weld pool.
3. **Switching hands:** In addition to applying the Walking the Cup technique, you will face an additional challenge—switching hands with the torch and filler rod.
   o   For the first half of the pipe, weld from 12 o'clock to 6 o'clock (passing through 3 o'clock) while holding the torch in your right hand and the filler rod in your left.
   o   Then, switch hands: hold the torch in your left hand and the filler rod in your right, welding the other half from 12 o'clock to 6 o'clock (passing through 9 o'clock).
   o   Mastering this hand-switching technique is essential for welding in various positions and ensuring accessibility to different joints.
4. **Angle control and progression:** Keep the torch at a constant angle while advancing around the pipe. If the angle changes, the bead quality may be compromised, leading to defects such as undercutting or lack of fusion. Ensure the tungsten is always focused on the weld pool at a consistent distance for optimal control.
5. **Filler rod feeding:** Introduce the filler rod at the edge of the molten pool, never in the centre, to avoid undesirable cooling and ensure proper fusion. As you progress around the pipe, hold the rod at a diagonal angle for better material deposition control. If the rod melts too quickly or deposits excess material, adjust its inclination to minimise direct contact with the weld pool.

6. **Practising hand-switching:** Changing hands is not easy, but it is a crucial skill for any welder. Practise this movement without striking an arc first to familiarise yourself with the hand motions. Once comfortable, apply the technique in actual welding. The goal is to maintain the same welding quality regardless of which hand is being used.

PERFORMING THE HOT PASS

- **Preparation:** After completing the first bead (root pass), conduct a quick visual inspection to ensure it is well-formed and free from defects such as porosity or lack of fusion.
- **Settings:** Maintain the same welding parameters, but adjust the current intensity slightly for the hot pass if necessary, increasing amperage to ensure proper fusion between the root pass and the new deposited material.
- **Execution:** Start the hot pass from the same starting point as the root pass. Again, apply the Walking the Cup technique, moving steadily around the pipe. It is crucial to maintain the weld pool just above the root bead to ensure full fusion between the layers. This step is essential for reinforcing the weld and eliminating any minor imperfections from the first pass.
- **Finalisation:** Maintain bead consistency throughout the process, finishing the hot pass with a smooth ramp-down and using the "rat tail" technique for a clean, well-blended finish.

FINAL CONSIDERATIONS

Stay calm and focus on your technique. Welding pipes in a horizontal position can be challenging, but with patience and practice, you will develop the skills necessary to create a uniform and high-quality weld bead. Remember that welding is both an art and a science—every practice session brings you closer to perfection.

THE CRACKED WATER JUG

In a small Tibetan village, an old man carried water from the river to his home using two jugs hanging from a pole over his shoulders. One of the jugs was cracked, while the other was in perfect condition. Every day, by the time he reached home, the perfect jug was full, while the cracked one only contained half the water.

One day, the cracked jug, filled with sadness, said to the old man:

"I feel useless. Because of my crack, I cannot fulfil my purpose. I am a failure."

The old man smiled and replied:

"Have you noticed the flowers growing along the path? I planted them knowing about your crack. Each day, as we carry the water, you unknowingly water those flowers. Thanks to you, I can enjoy their beauty and bring them to my home's altar. If you were not as you are, this path would be empty and dry."

Even your imperfections and challenges can be sources of learning, beauty, and contribution. The important thing is not just achieving perfection, but discovering the value of each step along the journey.

Want more information? Watch this exercise in action:

Facultad de Soldadura: www.youtube.com/watch?v=wvKpeNZnpb8

TIG WELDING ON CARBON STEEL

PRACTICE 12. PIPE TO PLATE IN VERTICAL UPWARD POSITION—PH/PF (2FR)

- **Base material:** Carbon steel plate 4″ x 4″ x 1/8″. Pipe 2″ Ø x 1–3/8″ x 1/8″.
- **Electrodes to use:** Choose from available options: 3/32-inch″ thoriated tungsten (WTh20 red), lanthanated (WLa 10/15 black or gold), ceriated (WCe 20 grey), or rare earth tungsten (E3 purple).
- **Nozzle:** No. 5 or No. 6 (5/16″ and 3/8″ diameter, respectively).
- **Filler rod:** 5/64″ or 3/32″ diameter, type ER 70S-6.
- **Number of beads:** Two beads (including a hot pass).
- **Current intensity:** 90–120 amps.
- **Gas flow rate:** 17–21cubic feet per hour (CFH).
- **Tungsten extension beyond the nozzle**: 1/8–5/32″.

PRELIMINARY ADJUSTMENTS

POSITIONING THE COUPON

Before starting, place the joint so that the **highest side of the pipe is at eye level**. This will allow you to keep a clear view of both the weld pool and the tungsten electrode throughout the welding process, which is crucial for optimal control (Figure 2.25).

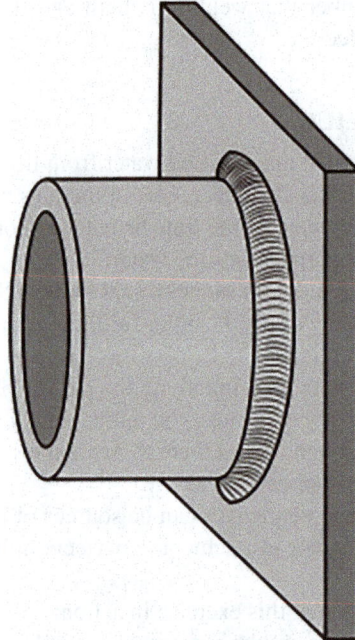

FIGURE 2.25 TIG welding of a tube to a plate in a vertical position using a fillet joint. The weld bead is visible around the base of the tube, securing it to the plate.

CHOOSING THE NOZZLE

Experimenting with different nozzle diameters can be beneficial for this practice.

- **A larger nozzle** (such as No. 6 or No. 7) provides wider shielding gas coverage, which is useful in situations where the weld pool may become unstable. However, a larger nozzle can make control more challenging in tight spaces.
- **A smaller nozzle** (such as No. 5) delivers a more concentrated gas flow, offering precise control in narrow areas, but requires greater dexterity to maintain gas protection across the entire weld zone.

I recommend trying both options and selecting the one that best suits your needs and comfort during welding. The key is to find a balance between weld pool control and shielding gas coverage.

PERFORMING THE WELD

Applying the Walking the Cup technique

1. **Head positioning and body rotation**
   In the vertical upward position, it is essential to keep your head between the torch and filler rod. This ensures that you always have a clear view of the weld pool and tungsten. As you progress in the weld, you must rotate around the pipe, maintaining the same attack angle, similar to the movement of a clock hand.
   o   Start at 6 o'clock and move upwards to 12 o'clock.
2. **Switching hands**
   Mastering hand-switching is crucial in this position. To facilitate the process:
   o   Weld from 6 o'clock to 10 o'clock (passing through 9 o'clock) holding the torch in your right hand and the filler rod in your left hand.
   o   Then switch hands to weld from 10 o'clock to 12 o'clock (torch in the left hand, filler rod in the right hand). This ensures better visibility of the upper part of the pipe without the torch obstructing your field of vision.
   o   For the other half of the pipe, reverse the process:
       –   From 6 o'clock to 4 o'clock with the torch in your left hand and the filler rod in your right hand.
       –   From 4 o'clock to 12 o'clock (passing through 3 o'clock) with the torch in your right hand and the filler rod in your left hand.
3. **Controlling the weld pool**
   In the vertical upward position, the weld pool tends to move downward due to gravity. To counteract this:
   o   Maintain a steady and controlled torch movement.
   o   The Walking the Cup technique or resting one or more fingers on the workpiece, using heat-resistant finger sleeves ("Fingers"), can provide greater stability and reduce hand tremors. This is particularly useful in this position, where precision control is critical.

FILLER ROD REEDING

Whether you decide to keep the filler rod free-floating or resting on the joint, it is crucial to always introduce it at the edge of the weld pool, never in the centre. This prevents undesirable cooling and ensures proper fusion.

In the vertical upward position, it is also recommended to feed the filler rod diagonally to better control material deposition. This helps prevent the weld pool from becoming unmanageable.

- If you notice the filler rod melting too quickly or excessive material buildup, adjust its angle to limit direct contact with the weld pool.

HOT PASS

After completing the first bead (root pass), it is time to apply the hot pass. This second bead reinforces the weld and corrects any defects from the root pass.

APPLYING THE HOT PASS

- Increase the amperage slightly (within the 90–120 amps range) to ensure proper weld pool penetration into the joint.
- Use the Walking the Cup technique to maintain a uniform and controlled movement, ensuring good fusion between the hot pass and the root bead.
- Maintain the same hand-switching strategy to guarantee a consistent weld quality around the entire circumference of the pipe.

FINAL CONSIDERATIONS

Welding in the vertical upward position requires greater control over both the weld pool and technique.

- First, practise the movements "dry" (without striking an arc) to get used to hand-switching and rotating around the pipe.
- Always maintain an ergonomic position to avoid fatigue and enhance precision.
- Ensure you use the proper PPE to stay protected throughout the process.

Now you have all the necessary information—remember: perseverance and constant practice are key to mastering this position and developing the dexterity needed to produce high-quality welds in any situation (Figure 2.26).

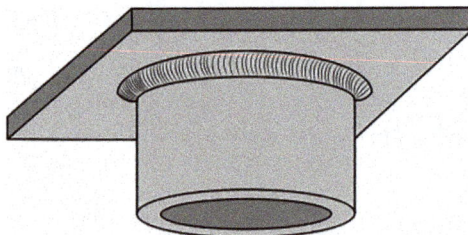

FIGURE 2.26   TIG welding of a tube to a plate in an overhead position using a fillet joint. The weld bead is visible, securing the tube to the plate.

TIG WELDING ON CARBON STEEL

PRACTICE 13. PIPE TO PLATE JOINT IN OVERHEAD POSITION—PD (4F)

- **Base material:** Carbon steel plate 4" x 4" x 1/8". Pipe 2" Ø x 1–3/8" x 1/8".
- **Electrodes to use:** Choose from available options: 3/32" thoriated tungsten (WTh20 red), lanthanated (WLa 10/15 black or gold), ceriated (WCe 20 grey), or rare earth tungsten (E3 purple).
- **Nozzle:** Free choice.
- **Filler rod:** 5/64" or 3/32" diameter, type ER 70S-6.
- **Number of beads:** Two beads (including a hot pass).
- **Current intensity:** 90–120 amps.
- **Gas flow rate:** 1 cubic foot per hour (CFH) per 0.04" of nozzle inner diameter.
- **Tungsten extension beyond the nozzle:** 1/8–5/32".

This practice focuses on welding a pipe to a plate in the overhead position. Since this exercise builds upon previous ones, we will emphasise the specific challenges of this position and how variations in nozzle selection affect the welding process.

EFFECT OF DIFFERENT NOZZLE DIAMETERS

When applying the Walking the Cup technique, the choice of nozzle diameter directly impacts the welding experience and bead progression:

- **Smaller nozzle (e.g., No. 5 5/16"):**
  o Concentrates the shielding gas more effectively, reducing the spread of the gas coverage.
  o Limits movement amplitude, meaning that even if you maintain the same motion cadence, the weld progresses more slowly, offering greater control over the weld pool.
  o Useful in tight spaces or when detailed control is required.
- **Larger nozzle (e.g., No. 8 1/2"):**
  o Expands the shielding gas coverage, increasing the movement range during Walking the Cup.
  o Allows for faster progression along the joint but may reduce precision over the weld pool.
  o Can be advantageous for covering larger sections efficiently, but may introduce the risk of losing control in overhead welding.

COMPARISON: STANDARD NOZZLE VS. GAS LENS

In previous exercises, you worked with ceramic nozzles of different diameters. Now is a good time to experiment by switching between a standard nozzle and a gas lens system.

- **Standard nozzle system:**
  o The **most common and versatile** option.
  o Gas flow **varies** depending on **nozzle size**, influencing how the weld pool is shielded.
- **Gas lens system:**
  o Provides a **more uniform and laminar gas flow**, enhancing gas coverage.
  o Particularly useful in **overhead welding**, where gas dispersion is common, exposing the weld to **oxidation risks**.
  o Using a **gas lens** instead of a **standard nozzle** may improve **stability** and **weld consistency**.

I strongly encourage you to test both systems during this practice. Switching between a standard nozzle and a gas lens will give you direct insight into how each affects weld quality and pool control. This experience will be valuable as you progress into more complex and demanding welding applications.

HAND-SWITCHING TECHNIQUE

As in the horizontal welding practice, it is essential to apply the hand-switching technique to achieve a uniform weld across the pipe:

- **Right side of the pipe (from 12 o'clock to 6 o'clock, passing through 3 o'clock)**
  - o Hold the torch in your right hand and the filler rod in your left hand.
- **Left side of the pipe (from 12 o'clock to 6 o'clock, passing through 9 o'clock)**
  - o Switch hands: Hold the torch in your left hand and the filler rod in your right hand.

This hand-switching method helps maintain an ergonomic posture and unobstructed visibility of the weld pool, preventing the torch from blocking your view.

FINAL REMINDER

We are now moving into more advanced aspects of TIG welding. Do not hesitate to experiment with different nozzle configurations and gas systems to find the optimal setup for your comfort and welding requirements.

- Always prioritise safety, ensuring the correct use of PPE.
- Maintain proper posture and ergonomics to prevent fatigue and errors.

Now, with all the necessary information at hand, remember:
Perseverance and continuous practice are the keys to mastering this position and developing the dexterity required for high-quality welds in any situation.

INSPECTION PROTOCOL FOR A CSWIP WI ON A 2″ PIPE COUPON WITH 1/8″ WALL THICKNESS WELDED TO A 1/8″ PLATE

In the context of a **welder qualification test**, the **CSWIP WI** follows a rigorous inspection protocol to ensure that the weld meets established quality and safety standards. This protocol is based on international standards, such as ISO 5817, which defines the acceptance criteria for welds, as well as the specifications of the applicable welding code for the given project.

1.   Preliminary visual inspection

Before performing any destructive or non-destructive testing (NDT), the CSWIP WI begins with a thorough visual inspection of the welded coupon. This initial examination focuses on the following aspects:

- **Weld bead uniformity:**
  - o The inspector assesses whether the weld bead is consistent around the entire circumference of the pipe and at the pipe-to-plate joint.

o   The weld profile must be regular, without undercuts, excessive reinforcement, or visible discontinuities.
- **Surface defects:**
  o   The inspector checks for visible defects, such as cracks, porosity, and lack of fusion.
  o   Special attention is given to open porosity and cracks, as these are critical defects that compromise the structural integrity of the weld.
- **Contamination and cleanliness:**
  o   The inspector verifies that the weld is free of contaminants, such as oxides or flux residues, which may indicate inadequate cleaning before or during the welding process.

2.   Dimensional measurements

The CSWIP WI then measures the weld dimensions to confirm that they comply with the **approved Welding Procedure Specification (WPS)**:

- **Weld width and reinforcement height:**
  o   The inspector checks whether the weld bead width and reinforcement height are within the acceptable range.
  o   Excessive filler material or undercut may result in coupon rejection.
- **Internal thickness:**
  o   If the weld sinks excessively into the inner wall of the pipe, this may indicate poor welding technique, such as slow travel speed or excessive heat input.

3.   Non-destructive testing (NDT)

Depending on the qualification test requirements, the CSWIP WI may perform additional NDT, such as:

- **Liquid penetrant testing (PT):**
  o   Detects fine cracks or porosity that are not visible to the naked eye.
- **Radiographic (RT) or ultrasonic testing (UT):**
  o   These methods assess the internal integrity of the weld and detect defects such as lack of fusion, slag inclusions, or internal porosity.

4.   Destructive testing (if required)

In some cases, destructive testing is necessary to evaluate the penetration and fusion of the welded joint:

- **Cut and macrographic examination:**
  o   A section of the coupon is cut and polished for macrographic analysis.
  o   The CSWIP WI examines root pass penetration and hot pass fusion, ensuring complete fusion between the pipe and plate, with no signs of lack of fusion or incomplete penetration.
- **Guided bend test (if applicable):**
  o   The coupon undergoes bending at specific angles to assess weld ductility and toughness.
  o   The inspector verifies that the weld remains intact and free of cracks or fractures.

5. Final evaluation and classification

The CSWIP WI evaluates all inspection results, including:

- Visual inspections
- Dimensional measurements
- Non-destructive and/or destructive testing

Based on the acceptance criteria defined in ISO 5817 or other applicable standards, the inspector determines whether the coupon meets the required quality standards.

- Classification:
  - o The weld is classified into quality levels, such as:
    - – B (High Quality)
    - – C (Medium Quality)
    - – D (Low Quality)
  - o For welder qualification, a minimum quality level of C is generally required.
  - o However, for critical projects, a level B weld may be mandatory.

6. Recording inspection results

Finally, the CSWIP WI documents all inspection results in a detailed report. This report serves as official documentation for the welder qualification and must include:

- All observations
- Dimensional measurements
- Results of NDT and destructive tests (if applicable)

CONCLUSION

The CSWIP WI inspection process is meticulous and thorough, ensuring that welds meet the highest standards of quality and safety.

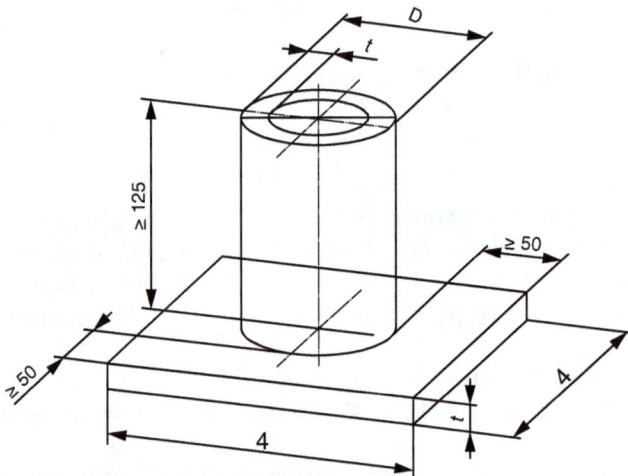

FIGURE 2.27   Dimensional diagram of a test specimen with a welded tube-to-plate joint. The minimum required dimensions for the welding test are indicated.

A welder's qualification depends not only on their technical skill but also on their ability to follow procedures and adhere to welding standards (Figure 2.27).

To increase the chances of passing the welder qualification test, the welder must:

✓ Train consistently
✓ Follow all recommended welding techniques
✓ Adhere to best practices in cleaning and preparation

WANT TO PRACTISE WITH A FULL-SIZE QUALIFICATION COUPON?

According to BS EN ISO 9606–1, the minimum dimensions (in inches) for a qualification test coupon are:

TIG WELDING ON CARBON STEEL

PRACTICE 14. WELDING OF 5″ PIPES IN HORIZONTAL POSITION—PC (2G)

- **Base material:** Two round carbon steel pipes, 5″ Ø (1″ = 2.54cm), 1/4″ wall thickness, and 1–9/16″ in length each (Figure 2.28).
- **Electrodes to use:** Choose from available options: 3/32″ thoriated tungsten (WTh20 red), lanthanated (WLa 10/15 black or gold), ceriated (WCe 20 grey), or rare earth tungsten (E3 purple).
- **Nozzle:** No. 5 or No. 6 (5/16″ and 3/8″ diameter, respectively).
- **Filler rod:** 5/64″ or 3/32″ diameter, type ER 70S-6.
- **Number of beads:** Two beads (including a hot pass).
- **Current intensity:** 90–120 amps.
- **Gas flow rate:** 17–21 cubic feet per hour (CFH).
- **Tungsten extension beyond the nozzle:** 1/8–5/32″.

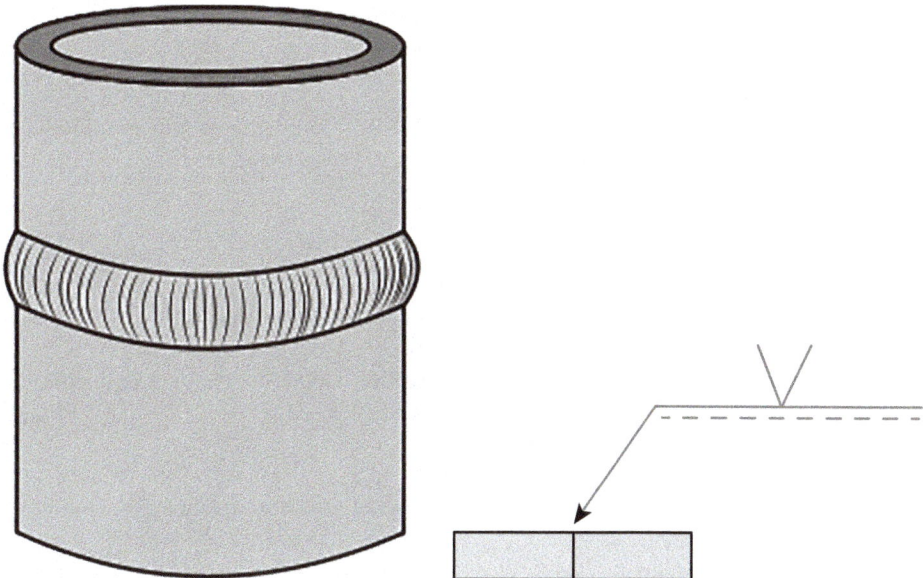

FIGURE 2.28   Circumferential welding in a pipe butt joint using the TIG process. The weld bead is visible around the joint between both pipe sections.

PREPARATION AND CLEANING

1. **Pipe preparation:**
   - o Ensure that the pipes are properly bevelled and that the root gap is appropriate.
   - o For this practice, a root gap of 3/32" is recommended when using a 3/32" filler rod, while a 1/8" gap is more suitable for a 1/8" filler rod.
   - o This gap ensures proper root penetration during welding.
2. **Cleaning the weld surfaces:**
   - o Clean the bevelled edges of the pipes and adjacent surfaces using a wire brush, sandpaper, or a specialised metal degreaser.
   - o Ensure that no traces of rust, oil, or dirt remain, as these contaminants can compromise weld quality.

TACKING AND POSITIONING

1. Tacking the joint:

- o Position the pipes in the horizontal welding position (PC 2G) and apply four evenly spaced tack welds around the circumference.
- o To maintain a root gap of 3/32" or 1/8", use a stripped welding electrode bent into a V shape as a temporary spacer.
- o This simple tool helps maintain a consistent gap until the tack welds are completed.
- o The tacks must be strong enough to hold the pipes in place during welding, but not excessively large, as this could interfere with the root pass flow.

2. Positioning the coupon:

- o Place the assembly in a comfortable and ergonomic position on your workbench.
- o Ensure that you can move around the pipes smoothly while welding, as you will need to adjust your body position to follow the bead progression and keep the tungsten in sight at all times.

PERFORMING THE WELD

1. Applying the Walking the Cup technique

- **Head Position and rotational movement:**
  - o Always keep your head between the torch and the filler rod to maintain an unobstructed view of the tungsten and weld pool.
  - o As you progress along the weld, rotate around the pipe while keeping a consistent torch angle.
  - o If at any moment you feel out of sync, it is better to stop and reposition yourself rather than letting the torch move ahead of your field of vision.
- **Controlling internal reinforcement:**
  - o The root pass should create a slight reinforcement on the inside of the pipe.
  - o If the weld appears concave instead of reinforced, this may indicate:
    - – Slow travel speed
    - – Excessive heat input
  - o Adjust your technique to avoid excess penetration and ensure a properly formed root bead.

2.   Welding options: Walking the Cup vs. finger support

- **Walking the Cup technique:**
  - o The most common method for pipe welding in the horizontal position.
  - o Keep the nozzle in contact with the pipe edges while moving forward with a side-to-side oscillation.
  - o This technique requires practice to maintain consistent motion and uniform fusion, especially in horizontal pipe welding (see Figure 2.29).
- **Finger support with heat-resistant sleeve (Fingers TM):**
  - o If you prefer more control, you can rest one or more fingers on the pipe surface while wearing a heat-resistant sleeve.
  - o This provides greater stability and reduces hand tremors.
  - o It is a valid alternative if you feel that Walking the Cup does not provide enough control in certain situations.

3. Filler rod diameter and its impact

- **3/32″ filler rod for a 3/32″ root gap:**
  - o Recommended for a narrower gap, allowing more controlled deposition and better root bead formation.
- **3/32″ or 1/8″ filler rod for a 1/8″ root gap**:
  - o With a wider root gap, a larger filler rod helps fill the joint properly without excessive reinforcement or incomplete fusion.

FINAL CONSIDERATIONS

- Maintaining consistent control over the weld pool is key to success in this exercise.
- Hand-switching and proper body rotation are essential for achieving a uniform weld and avoiding defects.
- If at any point you struggle with the process, stop, reposition yourself, and ensure that your field of vision is always centred on the tungsten and weld pool.
- Regular practice is the best way to perfect your technique. Each weld you complete is an opportunity to improve and develop greater skill.

FIGURE 2.29   Internal view of a circumferential pipe weld made using the TIG process. The weld bead is visible at the root of the joint.

THE APPRENTICE'S BRIDGE

In a remote valley, a raging river separated two villages. For years, the villagers dreamed of building a bridge, but the river was treacherous, and the challenge seemed impossible.

A young carpenter's apprentice, eager to prove his worth, volunteered to lead the construction. The elders looked at him with scepticism, but his determination was unwavering.

"I am neither the strongest nor the wisest, but I will give everything I have to succeed," he said.

With each plank he placed, he faced new obstacles:

- The materials didn't fit,
- The currents tried to wash him away,
- His own doubts haunted him.

However, every mistake taught him something valuable.

Day by day, his hands became calloused, his back bent, but his spirit remained strong.

Finally, after months of effort, the bridge was completed.

As the villagers crossed it for the first time, the apprentice, exhausted but proud, noticed something extraordinary:

- The marks of every difficulty were etched into the wood.
- They were not flaws, but memories of his effort, each one a testament to his growth.

An elder said to him:

"You did not just build a bridge for us, son. You built a bridge within yourself, linking your will to your ability. That is the true achievement."

Facing a difficult challenge, such as butt welding pipes, not only improves your technique—it also builds something within you.

Every weld you lay, every adjustment you make, is a step towards bridging who you are now with who you can become.

TIG WELDING ON CARBON STEEL

PRACTICE 15. WELDING OF 5″ PIPES IN VERTICAL UPWARD POSITION—PF (5G)

- **Base material:** Two round carbon steel pipes, 5″ Ø (1″ = 2.54cm), 1/4″ wall thickness, and 1–9/16″ in length each.
- **Electrodes to use:** Choose from available options: 3/32″ thoriated tungsten (WTh20 red), lanthanated (WLa 10/15 black or gold), ceriated (WCe 20 grey), or rare earth tungsten (E3 purple).
- **Nozzle:** No. 5 or No. 6 (5/16″ and 3/8″ diameter, respectively).
- **Filler rod:** 5/64″ or 3/32″ diameter, type ER 70S-6.
- **Number of beads:** Two beads (including a hot pass).
- **Current intensity:** 90–120 amps.
- **Gas flow rate:** 17–21 cubic feet per hour (CFH).
- **Tungsten extension beyond the nozzle:** 1/8–5/32″.

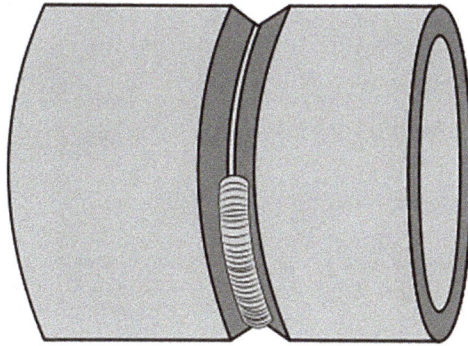

FIGURE 2.30 Circumferential welding in a V-groove butt joint on a pipe using the TIG process. The weld bead is visible on the outer part of the joint.

TUBE HEIGHT POSITIONING

Position the pipe so that the highest point is at eye level. This will facilitate access to both the lower and upper sections of the pipe (Figure 2.30).

Remember that in a welder qualification test, you cannot adjust the height of the coupon once the test has started.

Root pass

1. **Controlled upward movement**
   o   Maintain a steady and precise progression to prevent the molten metal from running down.
   o   The key is efficiency to avoid overheating the joint while progressing from 6 to 12 o'clock on both halves of the pipe.
2. **Torch angle and control**
   o   Keep a constant torch angle, ensuring that the tungsten remains focused on the edge of the weld pool.
   o   Slightly tilt the torch upwards to prevent the pool from sliding down.
3. **Pausing to remove tacks**
   o   Stop at the tack weld locations to grind them away with an angle grinder.
   o   This reduces the risk of defects in the weld bead.
4. **Internal reinforcement**
   o   Ensure that the root pass slightly protrudes on the inside of the pipe.
   o   If the weld appears concave instead of reinforced, adjust your travel speed or reduce the current to prevent excessive heat input.

Hot Pass

- **Applying the hot pass**
   o   The hot pass should be carried out using a slightly higher amperage (within the 90–120 amps range) to ensure good penetration and eliminate any minor defects left in the root pass.
   o   Maintain the same rhythm and control as in the root pass, but be especially careful to prevent the weld pool from sagging down.

ADDITIONAL TIPS

1.   Walking the Cup technique

- You can opt to use the Walking the Cup technique to keep the torch in continuous, controlled movement.
- This technique involves a gentle oscillating motion, supporting the ceramic cup against the pipe surface.
- This movement evenly distributes heat and provides precise control over the weld pool, reducing the risk of gravitational distortion.

2.   Filler rod positioning and progression

- The filler rod must always be positioned at the edge of the weld pool—never in the centre.
- This ensures a consistent material feed, avoiding excessive dilution and preventing burn-through caused by melting the joint edges without proper reinforcement.
- As you progress upwards, introduce the filler rod with a smooth and controlled motion.

3.   Filler rod feeding technique

- Use your thumb, index, and middle fingers to push the filler rod into the weld pool.
- The rod should slide smoothly between these fingers, with the thumb controlling the forward movement, while the other two fingers serve as a support point.
- This movement must be small and precise, allowing you to regulate the amount of material added at each moment.

4.   Finger support with heat-resistant sleeves (Fingers TM)

- If you find it difficult to maintain precise control with Walking the Cup, you can opt to rest your hand on the pipe using a heat-resistant sleeve (Fingers TM).
- This support helps to reduce hand tremors and provides better stability.
- Ensure that:
  - o   The torch and filler rod remain correctly aligned.
  - o   Your head stays positioned between the torch and the filler rod for a clear view of the tungsten and weld pool.

5.   Selecting the filler rod diameter

- For a 3/32" root gap → Use a 5/64" or 3/32" filler rod.
- For a 1/8" root gap → Use a 3/32" or 1/8" filler rod.
- Choosing the right filler rod diameter ensures better control over material deposition and prevents excessive buildup in the weld pool.

Final considerations

- Consistent control of the weld pool is crucial in this exercise.
- Hand-switching and body positioning are essential for achieving a uniform bead and preventing defects.
- If at any point the process becomes difficult, stop, reposition yourself, and refocus to ensure that your view remains centred on the tungsten and weld pool.

- Regular practice is the best way to master the technique. Every weld is an opportunity to improve and develop greater skill.

TIG WELDING ON CARBON STEEL

PRACTICE 16. WELDING OF 5″ PIPES IN 45° FIXED POSITION—HL-045 (6G)

- **Base material:** 2 round carbon steel pipes, 5″ Ø (1″ = 2.54cm), 1/4″ wall thickness, and 1–9/16″ in length each.
- **Electrodes to use:** Choose from available options: 3/32″ thoriated tungsten (WTh20 red), lanthanated (WLa 10/15 black or gold), ceriated (WCe 20 grey), or rare earth tungsten (E3 purple).
- **Nozzle:** No. 5 or No. 6 (5/16″ and 3/8″ diameter, respectively).
- **Filler rod:** 5/64″ or 3/32″ diameter, type ER 70S-6.
- **Number of beads:** Two beads (including a hot pass).
- **Current intensity:** 90–120 amps.
- **Gas flow rate:** 17–21 cubic feet per hour (CFH).
- **Tungsten extension beyond the nozzle:** 1/8–5/32″.

1. **Specific challenges of the 6G position**

   - The 6G position (HL-045) is one of the most challenging due to the pipe being fixed at 45° (Figure 2.31).
   - Since the pipe cannot be rotated, each weld segment must be executed at a different angle.
   - This requires greater control over the weld pool and precise coordination between the torch and the filler rod.

2. **Continuous movement and rotation**

   - As you progress through the weld, you must continuously adjust your position around the pipe to maintain a constant attack angle.

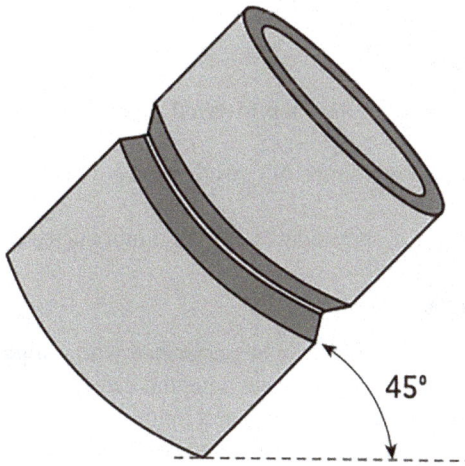

FIGURE 2.31   Bevelled butt joint on a pipe with a 45° angle for welding. The edge preparation is visible before the welding process.

- Unlike in previous positions, here you will need to constantly adapt your posture and the torch inclination to align with the pipe's angle.
- The key is to always keep the tungsten focused on the edge of the weld pool and to move smoothly without abrupt interruptions.

3. Using the Walking the Cup technique

- The Walking the Cup technique is particularly useful in this position.
- It helps maintain precise control over the weld pool as you rotate around the pipe.
- Keep the ceramic cup in contact with the pipe's surface and use a side-to-side oscillating motion to distribute heat evenly.
- Ensure that the cup remains in stable contact with the pipe, especially when welding in more awkward angles.

4. Hand-switching and posture adjustment

- In the 6G position, hand-switching is even more frequent and critical.
- You must change hands smoothly and at the right moment to maintain control over the weld pool.
- Plan ahead of time where you will switch hands, and adjust your body posture so you can make the transition without compromising weld quality.
- This exercise also tests your ability to maintain a consistent weld bead, regardless of which hand you are using.

5. Controlling the weld pool in critical areas

- In the 6G position, special attention must be given to the top and bottom sections of the pipe.
- Gravity plays a more significant role in these areas, causing molten metal to flow downward.
- Lower sections: Maintain a controlled travel speed and reduce amperage if the weld pool becomes difficult to control.
- Upper sections: Avoid excessive fusion by adjusting your travel speed and torch angle.

6. Removing tack welds

- As in previous welding positions, stop to grind away tack welds before welding over them.
- This is particularly important in the 6G position, where visibility and control are more limited.
- Removing tack welds significantly reduces the risk of introducing defects into the final weld.

7. Applying the hot pass

- The hot pass in the 6G position should be performed with a slightly higher amperage to ensure uniform penetration and eliminate potential defects.
- Maintain the Walking the Cup technique, and follow the recommended hand-switching strategy to ensure that the hot pass fuses seamlessly with the root pass.
- Maintain consistent travel speed and rotation around the pipe to achieve a high-quality weld (see Figure 2.32).

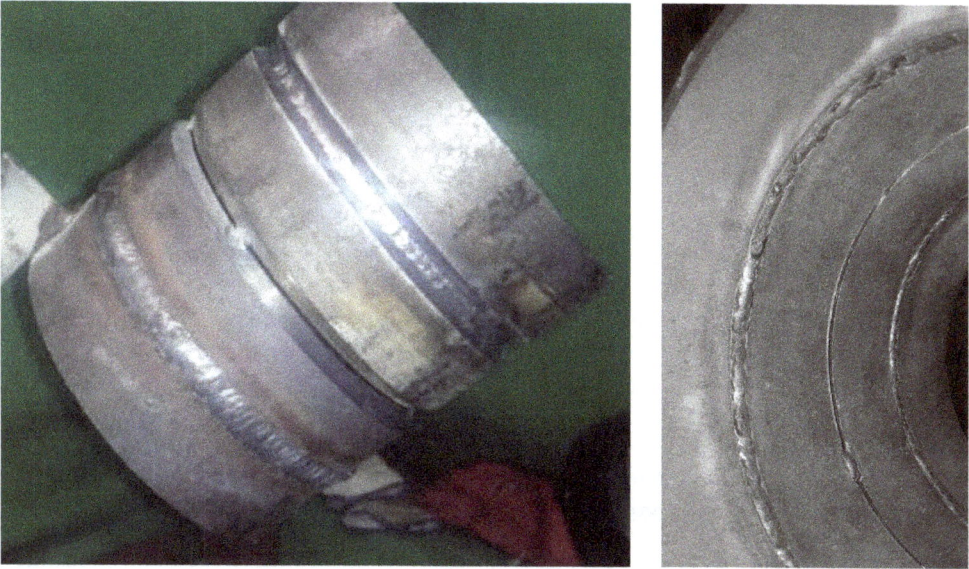

FIGURE 2.32   Pipe section with circumferential welds in a 45° bevelled joint. The deposited weld beads are visible at the connection between segments.

CONCLUSIONS

- Welding in the 6G position is one of the ultimate tests of a welder's skill.
- It requires:
  - ✓ **Precise control**
  - ✓ **Frequent hand-switching**
  - ✓ **Constant focus on posture and rotation**
- Practice and patience are essential to mastering this technique.
- If you follow the recommendations and maintain consistency, you will be able to produce a high-quality weld in this demanding position (Figure 2.32).

SUMMARY OF CHALLENGES AND CORRECTIVE ACTIONS FOR PIPE WELDING IN 5G AND 6G POSITIONS

The following list outlines the common difficulties encountered when welding carbon steel pipes in 5G and 6G positions, along with their corrective actions.

1.   Gravity and weld pool control

- ◇ **Weld pool sagging**
  - ☑ **Corrective action**: Maintain a slightly upward torch angle to counteract gravity.
  - ☑ Use the Walking the Cup technique or support your hand using a "finger" heat-resistant sleeve to stabilise the weld pool.
  - ☑ Maintain a steady pace and carefully control the filler rod feed to prevent the weld pool from becoming unmanageable.

- ◇ **Difficulty maintaining a controlled weld pool**
    - ☑ **Corrective action**: Reduce the amperage slightly if the weld pool becomes unstable.
    - ☑ Adjust the torch angle to direct heat precisely where needed, preventing excessive spread of the weld pool.

2. Visibility and torch angle control

- ◇ **Loss of tungsten visibility**
    - ☑ **Corrective action**: Keep your head positioned between the torch and the filler rod at all times.
    - ☑ If you lose sight of the tungsten, stop the weld, reposition yourself, and continue.
- ◇ **Difficulty maintaining a constant torch angle**
    - ☑ **Corrective action**: Rotate your body around the pipe, following the clock-hand technique to keep a steady torch angle.
    - ☑ If the angle starts shifting, stop and adjust your posture before proceeding.

3. Body positioning and posture adjustments

- ☑ **Corrective action**: Plan your weld path ahead of time, identifying hand-switching points in advance.
- ☑ Practise body rotation techniques to minimise strain and avoid uncomfortable postures.
- ☑ Reposition your body as needed to maintain comfort and control.

4. Tack welds

- ◇ **Removing tack welds**
    - ☑ **Corrective action:** Stop welding when reaching a tack weld and use an angle grinder with a cutting disc to remove it before continuing.
    - ☑ This reduces the risk of porosity and weld defects in the final joint.

5. Filler metal deposition

- ◇ **Excessive or insufficient filler metal**
    - ☑ **Corrective action:** Insert the filler rod at the edge of the weld pool, not in the centre, to prevent undesirable cooling and to control material deposition.
    - ☑ If the weld pool becomes difficult to control, adjust the filler rod angle to regulate deposition.
- ◇ **Difficulty maintaining the correct filler rod angle**
    - ☑ **Corrective action:** Practise dry runs to perfect the filler rod feeding angle.
    - ☑ Insert the filler rod diagonally to improve control.
    - ☑ If the rod melts too quickly, adjust its angle to minimise direct contact with the weld pool.

6. Heat input and pipe deformation

- ◇ **Heat accumulation**
    - ☑ **Corrective action:** Weld in short sections and alternate sides of the pipe to distribute heat evenly.
    - ☑ Allow the pipe to cool slightly between sections to prevent excessive heat build-up.
- ◇ **Pipe deformation**
    - ☑ **Corrective action:** Apply additional tack welds and use the staggered welding technique to minimise distortion.
    - ☑ If deformation is an issue, reduce the amperage and increase cooling time between passes.

7. **Work area access and mobility**

   ◇ **Restricted workspace**
      - ☑ **Corrective action:** Ensure that your work area is clear and that you have sufficient space to move around the pipe.
      - ☑ If space is limited, plan your rotation and movements in advance to avoid obstacles.
   ◇ **Uncomfortable posture**
      - ☑ **Corrective action:** Adjust the pipe height and position to maximise comfort.
      - ☑ Take short breaks to reposition yourself and reduce fatigue.
      - ☑ Practise dry-run movements to familiarise yourself with the necessary postures.

8. **Common weld defects**

   ◇ **Lack of fusion**
      - ☑ **Corrective action:** Maintain a proper torch angle and a consistent travel speed.
      - ☑ If fusion is insufficient, adjust the amperage and travel speed to improve penetration.
   ◇ **Porosity and slag inclusions**
      - ☑ **Corrective action:** Clean the weld surface thoroughly before starting.
      - ☑ Remove tack welds before welding over them.
      - ☑ If porosity or slag appear, stop welding, clean the area, and restart with a more precise technique.

WANT TO PRACTISE WITH A FULL-SIZE QUALIFICATION COUPON?

According to BS EN ISO 9606–1, the minimum dimensions (in inches) for a qualification test coupon are:

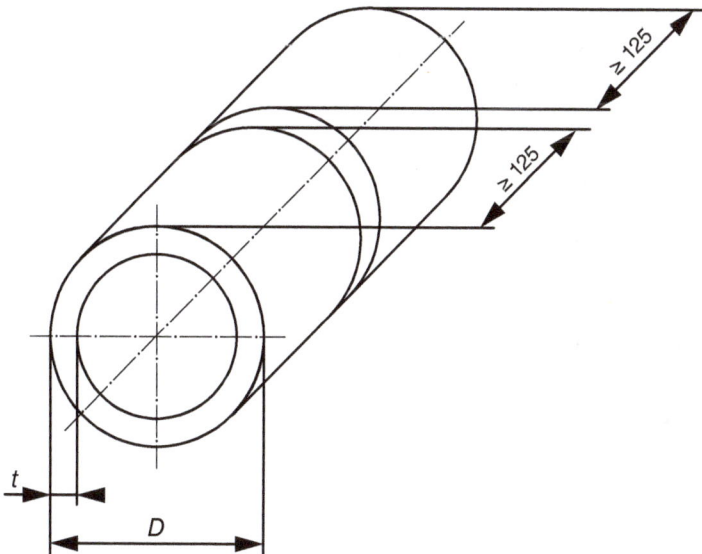

FIGURE 2.33 Dimensional diagram of a test specimen with a circumferential pipe weld. The minimum required dimensions for the welding test are indicated.

INSPECTION PROTOCOL FOR A CSWIP WELDING INSPECTOR (CSWIP WI) ON A 5″ PIPE-TO-PIPE COUPON

In the context of a welder qualification test, the inspection protocol for a 5″ pipe-to-pipe coupon with 1/4″ wall thickness, welded using the TIG process on carbon steel, follows a process very similar to the pipe-to-plate inspection protocol (Figure 2.33).

However, due to the specific characteristics of a larger diameter and thicker-walled pipe, certain additional aspects or variations may be required during inspection.

COMMON ASPECTS OF THE INSPECTION

1.   Visual inspection

◇   **Surface defects**
- ☑ The initial inspection is visual, checking for defects such as cracks, porosity, undercut, excessive reinforcement, or any visible discontinuity in both the root pass and hot pass.

◇   **Weld bead profile**
- ☑ The weld bead must be uniform, with a smooth transition between the weld and the base material.
- ☑ In thicker pipes, it is crucial to avoid undercut, as this weakens the weld integrity.

2.   Dimensional measurement

◇   **Weld bead width, height, and length**
- ☑ The weld dimensions must meet the approved Welding Procedure Specification (WPS), particularly in terms of reinforcement height.

◇   **Root gap verification**
- ☑ Ensuring that the initial root gap was correct is essential to prevent issues such as lack of penetration or root defects.

3.   Destructive and non-destructive testing (NDT)

◇   **Coupon cutting and fracture testing**
- ☑ If required by the inspection protocol, the coupon may be cut and fractured to assess penetration and fusion.
- ☑ The internal reinforcement should be adequate, with no lack of fusion, internal porosity, or slag inclusions.

◇   **Non-destructive testing (NDT)**
- ☑ Radiographic (RT) or Ultrasonic Testing (UT) can be used to evaluate the internal integrity of the weld.
- ☑ This is particularly important in thicker pipes, where internal defects can significantly affect structural integrity.

ADDITIONAL CONSIDERATIONS FOR LARGER PIPES

1.   Symmetry control

- ☑ Due to the larger pipe size, it is crucial to ensure that the weld is symmetrical around the entire circumference.
- ☑ Misalignment or uneven bead distribution can create internal stresses that compromise the structural integrity of the joint.

2. Distortion check

- ☑ In thicker pipes, distortion is less common but can still occur due to heat input, especially if multiple passes or combined welding processes were used.
- ☑ Verifying post-weld pipe alignment is essential.

3. Internal weld reinforcement assessment

- ☑ Internal reinforcement must be adequate to ensure a strong joint without weakening the base material.
- ☑ For 1/4″ (6mm) wall thickness pipes, internal reinforcement should be neither excessive nor insufficient to maintain mechanical integrity.

SUMMARY

- The inspection protocol remains largely similar to that of smaller pipes, but the increased complexity of a larger, thicker pipe requires additional focus on symmetry, distortion control, and NDT procedures.
- The applicable codes and standards, such as ISO 5817 or ASME Section IX, must be strictly followed during the inspection process.

INTRODUCTION TO TIG WELDING PRACTICES ON AUSTENITIC STAINLESS STEEL

Stainless steel is widely used in industry due to its corrosion resistance, which is primarily attributed to its chromium content.

However, to maintain these properties during welding, it is crucial to follow specific precautions and best practices.

PRECAUTIONS FOR HANDLING AND WELDING STAINLESS STEEL

☑ **Cutting and preparation**
- Use cutting discs and tools specifically designed for stainless steel (labelled "inox").
- This prevents cross-contamination with other materials, particularly carbon steel.

☑ **Storage**
- Store stainless steel separately from carbon steel and other alloys to prevent contamination.
- Dedicate specific tools for stainless steel work to avoid transferring particles from other materials.

☑ **Brushing**
- Use stainless steel wire brushes (usually silver in colour) for cleaning.
- Iron wire brushes (gold in colour) should not be used, as they can transfer contaminants that compromise weld quality.

☑ **Tack welding**
- At least five tack welds should be used to ensure proper alignment.
- Stainless steel has lower thermal conductivity than carbon steel, requiring a different tack welding sequence:
  - o First tack: Right end
  - o Second tack: Left end
  - o Third tack: Centre
  - o Fourth and fifth tacks: Equidistant between centre and ends

☑ **Cleaning**
  - Before welding, clean surfaces with an appropriate degreaser, such as acetone, to prevent contamination that could affect weld quality.

☑ **Backside protection**
  - When welding stainless steel with **TIG or MAG**, protect the **backside of the weld** to prevent **oxide formation**.
  - Use **copper backing bars** or **ceramic backing tape** to shield the **root pass**.
  - Without **proper protection**, **chromium can oxidise**, forming **"black crystals"**, which indicate **contamination**.

GAS PURGING IN STAINLESS STEEL PIPE WELDING

Gas purging is essential when welding stainless steel pipes, as it protects the internal surface from exposure to oxygen, preventing oxidation and defects.

◇ **Applying the gas purge**
  - ☑ Purging is carried out by introducing an inert gas (typically argon) inside the pipe before and during welding.
  - ☑ The pipe ends must be sealed, leaving small vent holes to allow air to escape.

◇ **Purging time and gas flow rate**
  - ☑ The required purging time depends on the pipe diameter, length, and gas flow rate.
  - ☑ A typical gas flow rate is 21–32 cubic feet per hour (10–15L/min), maintained until the pipe interior is fully inert.

◇ **Detecting residual oxygen**
  - ☑ Oxygen sensors can measure residual oxygen levels inside the pipe.
  - ☑ A level of 20 ppm or lower is generally considered acceptable before starting the weld, as it minimises oxidation risks.

FORMATION OF CHROMIUM OXIDES AND CHROMIUM CARBIDES

When welding stainless steel, it is essential to prevent the formation of chromium oxides and chromium carbides, as both can compromise corrosion resistance.

1. Chromium oxides

◇ **Formation**
  - ☑ Chromium oxides form when stainless steel is exposed to oxygen at high temperatures, such as during welding.
  - ☑ This appears as discolouration or black oxidation ("black cauliflower") on the weld surface.

◇ **Effect**
  - ☑ A thin layer of chromium oxide is protective (part of stainless steel's passive layer), but excess oxidation weakens this layer, reducing corrosion resistance.

◇ **Prevention**
  - ☑ Use proper gas purging to protect both the front and back sides of the weld.
  - ☑ Minimise exposure to high temperatures and maintain a short arc length with a steady travel speed to reduce oxidation.

2. Chromium carbides

◇ **Formation**
- ☑ Chromium carbides form when chromium reacts with carbon in stainless steel at temperatures between 450°C and 850°C (842°F–1562°F).

◇ **Effect**
- ☑ Chromium carbide precipitation reduces the amount of chromium available to form the passive layer, leading to intergranular corrosion.

◇ **Prevention**
- ☑ Use low-carbon stainless steel (304L or 316L) to minimise carbide precipitation.
- ☑ Control welding heat input and cool the material quickly after welding.

TYPES OF STAINLESS STEELS AND THEIR WELDABILITY

Stainless steel is classified into several types based on its microstructure and composition. The most common types are:

1. Austenitic stainless steels

- ☑ The most widely used in industry.
- ☑ Highly corrosion-resistant and non-magnetic.
- ☑ Example: 304L stainless steel, typically welded using ER308L filler material.
- ☑ Excellent weldability and low susceptibility to cracking during welding.

2. Ferritic stainless steels

- ☑ Magnetic but more prone to pitting corrosion and stress corrosion cracking.
- ☑ Used in applications where corrosion resistance is less critical, such as automotive components.
- ☑ Moderate weldability, requiring proper heat control to prevent grain growth.

3. Martensitic stainless steels

- ☑ High strength and hardness, but less corrosion-resistant than other stainless steels.
- ☑ Used in applications requiring high strength, such as blades and tools.
- ☑ Welding requires preheating and post-weld heat treatment to prevent cracking.

4. Austenitic-ferritic (duplex) stainless steels

- ☑ Combine the properties of austenitic and ferritic stainless steels.
- ☑ Higher mechanical strength and superior corrosion resistance.
- ☑ Common in marine and petrochemical applications.
- ☑ Typically welded using ER2209 filler material to maintain the duplex microstructure.

SKILL EXPANSION: WELDING 2″ PIPES WITH 1/8″ WALL THICKNESS

Once you have mastered the previous TIG welding practices on 5″ pipes with 1/4″ wall thickness, we encourage you to take the next step in your training by familiarising yourself with welding smaller-diameter, thinner-walled pipes.

Specifically, working with 2″ diameter pipes with a 1/8″ (3mm) wall thickness will allow you to expand your skillset and prepare for certification in small-diameter pipe welding—a highly valued competency in the industrial sector.

SPECIFIC CHARACTERISTICS OF WELDING 2″ PIPES WITH 1/8″ WALL THICKNESS

1. Full TIG welding process

☑ Unlike 5″ pipes with 1/4″ wall thickness, which are typically welded using a combined process (TIG root pass followed by stick welding for fill and cap passes), 2″ pipes with 1/8″ wall thickness can be fully welded using TIG.

☑ This means that both the root pass and hot pass can be performed with TIG, achieving a high-quality and precision joint.

2. Adjustments and techniques

☑ Due to the thinner wall thickness, it is crucial to adjust the amperage and weld pool control to prevent burn-through or pipe distortion.

☑ Welding 2″ pipes with TIG requires even greater control, particularly when advancing the root pass.

☑ The Walking the Cup technique remains highly effective in maintaining constant control over the weld pool, but it will require a lighter and more precise touch.

3. Challenges and benefits

☑ Welding smaller-diameter and thinner-walled pipes presents unique challenges, such as:
- The metal overheating quickly.
- The need for a steady rhythm to avoid root burn-through.
- ☑ However, mastering this skill will allow you to work on a wider range of industrial applications, particularly in:
- Confined-space welding.
- High-pressure piping systems.

Preparing for certification

☑ Once you become confident welding 2″ pipes with 1/8″ wall thickness, you will be well-prepared to pursue certifications in small-diameter pipe welding.

☑ These certifications are highly sought after in industries such as:
- Chemical and petrochemical
- Food processing
- Pharmaceutical sectors

☑ 2″ pipe welding is an excellent step towards refining your technique and expanding your career opportunities in specialised welding projects.

☑ Take full advantage of this training opportunity to strengthen your professional profile and unlock new career prospects in the welding industry.

TIG WELDING ON STAINLESS STEEL E-304L

PRACTICE 17. BUTT WELDING OF THIN-WALLED PIPES IN A HORIZONTAL ROTATING POSITION—PA (1G)

- **Base material:** 2" Ø x 3–15/16" (100mm) long x 1/16" (1.5mm) wall thickness austenitic stainless steel 304L.
- **Electrodes to use:** Choose from available options: 3/32" thoriated tungsten (WTh20 red), lanthanated (WLa 10/15 black or gold), ceriated (WCe 20 grey), or rare earth tungsten (E3 purple).
- **Nozzle:** No. 5 or No. 6 (5/16" and 3/8" diameter, respectively).
- **Filler rod:** None (Autogenous weld).
- **Number of beads:** One bead.
- **Current intensity:** 45–70 amps.
- **Gas flow rate:** 17–21 cubic feet per hour (CFH).
- **Tungsten extension beyond the nozzle:** 1/8–5/32".

1. Cleaning protocol

Material cleaning is essential before starting any welding process to avoid defects and ensure high-quality welding (Figure 2.34).

☑ **External pipe cleaning**
- Use a clean cloth soaked in acetone to wipe the pipe surface around the joint.
- Acetone effectively removes grease or dirt that could compromise weld quality.

☑ **Internal pipe cleaning**
- Ensure the inside of the pipe is contaminant-free, as this could affect gas purging quality and compromise the weld's root side.
- Use a stainless steel wire brush and, if necessary, clean the inside with acetone.

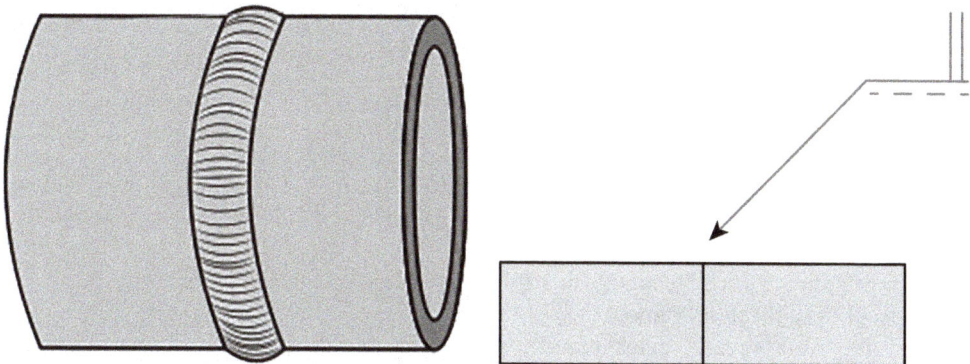

FIGURE 2.34   Circumferential welding in a pipe butt joint using the TIG process. The weld bead fully covers the joint between the two pipe sections.

2.   Pipe assembly and tack welding

Proper assembly and tack welding are crucial for maintaining precise alignment during welding.

☑ **Assembly**
- Align the pipes carefully in a stable fixture to ensure they remain in the correct horizontal position (PA/1G).
- Ensure zero root gap, as no filler rod will be used.
- A slight reinforcement on the inner weld face is desired.

☑ **Tack welding**
- Apply at least four tack welds, evenly distributed around the pipe's circumference.
- Stainless steel has lower thermal conductivity, meaning heat concentrates more around the weld pool, increasing the risk of distortion.

3.   Gas purging for the weld root

Gas purging is essential when welding stainless steel to prevent oxidation and protect the pipe's internal surface.

☑ **Applying the gas purge**
- With the pipe ends sealed, introduce argon gas inside at a constant flow of approximately 21 cubic feet per hour (10L/min).
- Maintain the purge throughout the welding process, ensuring the interior remains fully inert.
- Drill a small vent hole at the highest point of the opposite pipe end to allow trapped air to escape.

☑ **Residual air detection**
- Use an oxygen monitor to ensure oxygen levels inside the pipe are below 20 ppm before welding.
- This prevents chromium oxide formation on the weld root face.

4.   Welding execution

This practice involves autogenous welding (without filler material), requiring precise weld pool control to form a root bead with slight reinforcement.

WALKING THE CUP TECHNIQUE

☑ **Starting the technique**
- Position the torch at the starting point of the weld, around the 3 o'clock position on the pipe.
- Maintain a torch inclination of 10–15 degrees towards the direction of travel.

☑ **Oscillating torch movement**
- The Walking the Cup technique involves a side-to-side oscillating motion, with the ceramic cup resting on the pipe surface.
- This movement evenly distributes heat and ensures controlled fusion.
- Imagine "drawing" small overlapping arcs with the torch to achieve continuous fusion without overheating.

☑ **Pacing and consistency**
  - A steady rhythm is essential:
    - o **Too fast = Incomplete fusion or an irregular bead.**
    - o **Too slow = Excessive heat input, distortion, or burn-through.**
  - If the torch slips on the pipe surface, slightly reduce the inclination angle to increase ceramic cup friction for better control.

☑ **Avoiding unwanted stops**
  - Do not hold the torch in one place, as this can cause excessive heat concentration.
  - Maintain a continuous and fluid oscillation, applying moderate forward pressure to match the fusion rate.

☑ **Advancing and weld pool control**
  - Monitor the weld pool constantly to ensure it remains controlled.
  - Aim for a small, consistent weld pool that produces a slight reinforcement on the weld root.
  - If the weld pool becomes unstable, reduce the current or adjust travel speed.

☑ **Quarter-weld progression**
  - Weld the first quarter of the pipe (3 o'clock to 12 o'clock), then stop and inspect the bead.
  - Rotate the pipe 180° and weld the opposite quarter.
  - Repeat the same process for the remaining two quarters.

FINAL CONSIDERATIONS

☑ **Temperature and cooling**
  - Monitor the pipe temperature during welding.
  - If it overheats, allow a short cooling period before continuing.
  - Excessive heat can compromise weld integrity.

☑ **Leak test**
  - Once the weld is complete, perform a leak test to ensure no porosity or cracks compromise the joint's airtightness.

☑ **Visual inspection**
  - Check that the bead has uniform reinforcement and no excessive discolouration (a sign of chromium oxidation).

SUMMARY

✓ **Zero root gap setup for autogenous welding**.
✓ **Walking the Cup technique for controlled fusion**.
✓ **Consistent purge gas flow to prevent oxidation**.
✓ **Steady rhythm and heat control to avoid burn-through**.
✓ **Quarter-weld progression to ensure symmetry**.

This exercise **refines your TIG welding precision** on **thin-walled stainless steel pipes**, preparing you for **critical piping applications** in industries such as **pharmaceutical, food processing, and aerospace**.

UPCOMING CHALLENGES: WELDING IN PC, 5G, AND 6G POSITIONS

Once you have successfully completed welding in the horizontal position, we encourage you to take your skills to the next level by performing the same exercise in more challenging positions:

✓ Cornice (PC/2G)
✓ 5G (Fixed Horizontal)
✓ 6G (Inclined 45°—Fixed Pipe)

These positions will require greater control over the weld pool, torch movement, and gas purging technique.

The following section covers key aspects to consider when tackling these new positions:

1. Weld pool control

☑ In cornice, 5G, and 6G positions, weld pool control becomes more critical due to gravity's effect.
☑ It is essential to maintain a consistent torch angle and adjust amperage to prevent weld pool instability that could lead to defects in the bead.
  - **PC (2G)/5G position:**
    o Work in sections and maintain steady torch movement.
    o Pay close attention to the angle to prevent weld pool sagging.
  - **6G position:**
    o This is the ultimate challenge.
    o You must rotate your body around the pipe to maintain constant weld pool control as you progress from the bottom to the top.

2. Walking the Cup technique

☑ The Walking the Cup technique remains your best ally, but it will be more demanding in these positions.
☑ Keep an oscillating, continuous movement, ensuring the ceramic cup is always in contact with the pipe surface, particularly in vertical positions (5G and 6G).
  - If you feel you are losing control:
    o Stop
    o Adjust your posture, amperage, or torch inclination
    o Continue when you feel secure

A slip or pause in these positions can result in excessive heat buildup or an irregular bead formation.

3. Gas purging application

☑ In 304L austenitic stainless steel welding, gas purging is essential to prevent oxide formation and maintain chromium integrity.
  - **Purge flow rate and time**
    o Adjust the purge gas flow (typically 17–21 cubic feet per hour/8–10L/min).
    o Allow enough time to fully inert the pipe interior.
    o Use an oxygen sensor (PPM detector) to confirm oxygen levels below 20 ppm before starting the weld.

- **Continuous monitoring**
  - o In complex positions like 5G and 6G, movement can disrupt the purge flow.
  - o Ensure purge stability throughout the entire welding process to prevent internal defects.

4. Hand switching and body control

☑ In 5G and 6G positions, switching hands and rotating your body to maintain weld pool control is crucial.
  - Plan your welding path in advance, anticipating hand changes and positional adjustments.
  - Practise these movements in a dry run before starting the actual weld.
  - Keep your head positioned ahead of the torch and weld pool for better visibility and control.

5. Heat management

☑ Stainless steel has low thermal conductivity, meaning heat accumulates quickly in the weld zone.
  - Work in short sections and allow cooling intervals if needed.
  - Avoid overheating the top section in 6G, where heat buildup is most likely.

MASTERING THESE ADVANCED TECHNIQUES

By mastering these advanced welding positions, you will be one step closer to becoming a highly skilled welder capable of handling the industry's most complex challenges.

Keep practising, refining your technique, and pushing your limits!

CSWIP WELDING INSPECTOR (CSWIP WI) INSPECTION PROTOCOL FOR A 304L AUSTENITIC STAINLESS STEEL PIPE WITH A 1/16″ (1.5MM) WALL THICKNESS

The inspection protocol for welding a 304L austenitic stainless steel pipe with a 1/16″ (1.5mm) wall thickness follows the same fundamental steps as for a 5″ (127mm) x 1/4″ (6mm) carbon steel pipe.

However, there are key differences due to:

✓ Material properties
✓ Thin-wall characteristics

1. Visual inspection

Similarities

☑ Visual inspection remains essential to detect:
  - Surface defects (cracks, porosity, undercut, discontinuities)
  - Uniform bead profile with smooth transition between the bead and base material

Differences

☑ With 304L stainless steel (1/16″ wall thickness), it is critical to:
  • Check for oxidation or discolouration in the weld area.
  • Avoid heat tint colours such as blue, purple, or brown, which indicate excessive heat exposure and oxidation, potentially compromising corrosion resistance.
  • Prevent edge burn-through, a common issue in thin-walled stainless steel due to excessive heat input.

2. Dimensional measurement

Similarities

☑ Weld bead dimensions must comply with WPS specifications.

Differences

☑ Critical evaluations for thin-wall stainless steel:
  • Initial root gap assessment is vital to prevent excessive heat input or lack of penetration.
  • Minimal reinforcement on both root and cap passes to prevent structural weakening due to high distortion risk.

3. Destructive and non-destructive testing (NDT)

Similarities

☑ The coupon may be cut and fractured to assess:
  • Penetration and fusion
  • Absence of internal defects
  • ☑Radiographic Testing (RT) or Ultrasonic Testing (UT) can assess weld integrity.

Differences

☑ Increased risk of distortion and overheating due to:
  • Thinner material thickness
  • Higher susceptibility to thermal stress
☑ Special attention should be given to:
  • Structural weakening or deformation in destructive and non-destructive tests
  • Corrosion resistance verification, ensuring no contamination from other metals (to prevent galvanic corrosion)

4. Additional considerations for 1/16″ (1.5mm) wall thickness pipes

Heat-affected zone (HAZ) control

☑ Due to thin material thickness, heat concentration in the HAZ must be minimised.
☑ The welder must:
  • Use low heat input techniques such as pulsed TIG welding.
  • Work in short segments, allowing for intermittent cooling.

Gas purging

- ☑ Maintaining an effective gas purge is essential to prevent oxidation on the root pass.
- ☑ The purge must continue until the weld cools sufficiently to prevent post-weld oxidation.

SUMMARY

The CSWIP WI inspection protocol for 1/16″ (1.5mm) 304L stainless steel pipe welding follows the same fundamental steps as for a 5″ (127mm) x 1/4″ (6mm) carbon steel pipe.

However, it adapts to the unique properties of stainless steel, focusing on:

- ✓ Preventing oxidation
- ✓ Minimising heat input
- ✓ Ensuring stricter quality control due to thin-wall sensitivity

READ THIS BEFORE CONTINUING

In the following exercises, we will focus on TIG welding of aluminium, specifically using 5086 series aluminium alloy. This magnesium-based alloy is widely used in the aerospace and automotive industries due to its lightweight properties and high strength.

SPECIFIC CHARACTERISTICS OF ALUMINIUM AND ITS WELDING PROCESS

Aluminium presents several challenges compared to carbon steel or stainless steel.

One critical factor is the presence of alumina (aluminium oxide) on its surface.

- Alumina has a melting point of approximately 3,632°F (2,000°C), which is significantly higher than that of aluminium itself, which melts at 1,220°F (660°C).
- Alumina is ceramic in nature, making it refractory and insulating, hindering the welding process unless it is properly removed beforehand.

ALTERNATING CURRENT (AC) AND ITS EFFECT ON WELDING

- ☑ TIG welding of aluminium requires Alternating Current (AC), which provides a "cleaning effect" on alumina during the welding process.
  - The positive cycle of AC removes alumina.
  - The negative cycle allows arc penetration into the material.
- ☑ This cleaning action is essential to ensure a high-quality weld free from oxide inclusions.

THERMAL CONDUCTIVITY OF ALUMINIUM

- ☑ Aluminium has high thermal conductivity, meaning it dissipates heat rapidly.
- ☑ This requires higher amperage to maintain a stable weld pool.

However, this same property also means aluminium heats up quickly, which can negatively affect the HAZ and mechanical properties.

To mitigate this risk, it is recommended to:

✓ Weld in short segments, allowing the material to cool between passes.
✓ Use pulsed arc welding to control heat input more effectively.

IMPORTANCE OF CLEANING AND USING NOBLE GASES

☑ Before welding, it is essential to clean aluminium surfaces with a suitable degreaser or acetone.
   ✓ Acetone evaporates quickly and leaves no residue.
   ✓ Aluminium requires a higher shielding gas flow rate to prevent contamination.
☑ Use noble inert gases, such as:
   ✓ Pure argon
   ✓ Argon-helium mixtures

This ensures the weld pool remains uncontaminated.

SAFETY CONSIDERATIONS WHEN HANDLING ALUMINIUM

☑ Alumina dust is both flammable and hazardous if inhaled.
   ✓ Keep the work area clean.
   ✓ Wear fire-resistant clothing.
   ✓ Use an appropriate respirator mask.

Although AC removes alumina, pre-welding oxide removal remains essential to achieve defect-free welds.

HEAT CONTROL IN THE HAZ

Due to aluminium's high thermal conductivity, temperature control in the HAZ is critical for preserving mechanical properties.

Recommended strategies to minimise HAZ alterations:

✓ Use pulsed arc welding.
✓ Weld in short passes, allowing cooling between passes.
✓ Maintain controlled heat input to ensure material integrity.

Description of pulsed arc and square wave balance: What they are and why everyone confuses them.

When discussing TIG welding, you will often hear terms like "pulsed arc" and "square wave balance". While they may sound complicated, I will explain them in a simple way so you can understand what they are and why they are often confused.

1. Pulsed arc

What is pulsed arc?

☑ Pulsed arc is a TIG welding technique that rapidly alternates between a high and low current level while welding.
☑ Imagine you are accelerating and decelerating a car continuously, but in a controlled manner.

Why use it?

✓ Better heat control, particularly useful when welding thin materials that can warp or burn if exposed to excessive heat.

✓ Reduces the risk of burn-through and improves control over the size and shape of the weld pool.

Example:

◊ Imagine welding an aluminium can. If you use too much current at once, you could melt through it. Pulsed arc welding prevents this by giving more control over the heat input.

2. Square wave balance

What is square wave balance?

☑ In TIG welding with AC, the current constantly changes direction (from positive to negative and vice versa).

☑ Square wave balance controls the time spent in each direction.

☑ It is primarily used when welding aluminium.

Why is it important?

✓ When the current is in the positive half-cycle (electrode positive—EP), it helps remove aluminium oxide (alumina).

✓ When the current is in the negative half-cycle (electrode negative—EN), it focuses heat on the material, allowing it to melt and form the weld pool.

✓ Adjusting square wave balance helps to find the right ratio between cleaning and penetration.

Example:

◊ Think of it as washing a car with a hose.
  • You need enough pressure to remove the dirt (electrode positive– cleaning).
  • But not so much pressure that it damages the paint (electrode negative—penetration).

3. Why are they confused?

Similarities and confusion

☑ Both pulsed arc and square wave balance involve adjusting the current during welding.

☑ Both affect heat control, but in different ways.

How to differentiate them?

✓ **Pulsed arc:** Controls the "rhythm" of welding by alternating between more and less heat.

✓ **Square wave balance:** Controls how current is distributed between cleaning and penetration in aluminium welding.

EFFECT OF VARYING PULSE FREQUENCY IN PULSED ARC WELDING

Pulse frequency refers to how fast the current switches between high and low levels.

Imagine you are switching a light on and off repeatedly—pulse frequency is the number of times you do this per second.

1. Low pulse frequency

What happens when pulse frequency is low?

- ☑ The high-current phase lasts longer before switching to low current.
- ☑ This results in a slower welding pace and greater control over the weld pool.

When Is It Useful?

- ✓ Ideal for thicker materials where heat needs to penetrate deeper.
- ✓ Useful for beginners, as it provides more time to react to the weld pool.

Example:

- ◇ Painting carefully with slow strokes to stay within the lines for detailed work.

2. High pulse frequency

What happens when pulse frequency is high?

- ☑ The current switches rapidly between high and low.
- ☑ Heat is applied in short bursts, which keeps the weld pool smaller and more controlled.

When is it useful?

- ✓ Best for thin materials that are prone to warping or burn-through.
- ✓ Helps to increase travel speed while maintaining control.

Example:

- ◇ Like using a fine brush to quickly paint many small areas while still maintaining precision.

3. How to choose the right pulse frequency?

Based on material thickness:

- ✓ Thick materials → Low frequency for better heat penetration.
- ✓ Thin materials → High frequency to prevent overheating or warping.

Based on comfort level:

- ✓ If losing control of the weld pool → Lower frequency.
- ✓ If welding too slowly or struggling with precision → Increase frequency.
  - ☑ Adjusting pulse frequency correctly improves both weld quality and control.

EFFECT OF VARYING SQUARE WAVE BALANCE

In AC TIG welding, especially for aluminium, adjusting square wave balance is critical for controlling arc behaviour and achieving high-quality welds.

1. What is square wave balance?

Understanding the Wave

- ☑ In AC welding, the current alternates between positive and negative cycles.
- ☑ The wave balance setting adjusts how much time is spent in each half-cycle:
  - ✓ **Electrode positive (EP) → More cleaning**
  - ✓ **Electrode negative (EN) → More penetration**

2. Balance towards electrode positive (More cleaning)

What happens?

- ☑ The current spends more time in the positive cycle (EP).
- ☑ Increases oxide removal, ensuring a cleaner weld surface.
- ☑ However, it also heats the tungsten more, causing faster tungsten wear.

When is it useful?

- ✓ Ideal when welding aluminium with heavy oxidation.
- ✓ Ensures a clean weld before fusion begins.

Example:

- ◇ Scrubbing a surface before painting—you spend more time cleaning to remove all dirt before starting.

3. Balance towards electrode negative (More penetration)

What happens?

- ☑ The current spends more time in the negative cycle (EN).
- ☑ More heat goes into the workpiece, increasing penetration.
- ☑ However, less time is spent on cleaning, so oxide removal is reduced.

When is it useful?

- ✓ Best for pre-cleaned aluminium where deep penetration is needed.
- ✓ Ensures stronger welds with better fusion.

Example:

- ◇ Driving a nail deeper into wood—focusing more force into penetration rather than surface cleaning.

4. *How to choose the right square wave balance?*

- ✓ **For more cleaning:**
  - → Increase electrode positive (EP) time.
- ✓ **For more penetration:**
  - → Increase electrode negative (EN) time.
- ☑ **Balancing both factors ensures a high-quality aluminium weld.**

Summary

✓ Pulsed Arc = Adjusts heat input rhythm for better temperature control.
✓ Square Wave Balance = Adjusts current distribution for optimal cleaning vs. penetration in aluminium welding.

Mastering both settings will improve your welding results and overall precision.

FILLER MATERIAL: 5356 VS 4043—WHAT TO EXPECT AND HOW TO CHOOSE TO AVOID POROSITY AND CRACKS

When welding aluminium, selecting the right filler material is essential to achieving a good weld. The two most common filler materials are **5356 and 4043**. In the following section, I'll explain in simple terms what you can expect from each and how to choose the right one to avoid issues like porosity and cracks.

1. Alloy 5356

- **What is it?**
  o 5356 is a filler material that contains magnesium. It is known for being strong and highly resistant to corrosion, particularly in humid or marine environments.
- **When to use it:**
  o If you are welding aluminium alloys that also contain magnesium (such as the 5xxx series), 5356 is a great choice.
  o It is ideal if the part you are welding will be anodised, as it maintains a better colour finish.
- **Advantages and disadvantages:**
  o ✓ **Advantage:** Less porosity in the weld.
  o ✗ **Disadvantage:** Can be slightly more challenging to work with compared to 4043 in some cases.

2. Alloy 4043

- **What is it?**
  o 4043 is a filler material that contains silicon. It has a lower melting point, meaning it melts more easily and provides better flow during welding.
- **When to use it:**
  o It is suitable for welding aluminium alloys that contain silicon (such as the 6xxx series) and when you need a smooth, flowing weld to fill gaps effectively.
  o It is less suitable for anodising, as it can cause colour variation in the finished part.
- **Advantages and disadvantages:**
  o ✔ **Advantage:** Lower risk of cracking, especially when welding crack-sensitive alloys.
  o ✗ **Disadvantage:** Can be more prone to porosity if the welding process is not properly controlled.

3. How to choose between 5356 and 4043

- **To avoid porosity:**
  o If corrosion resistance and a good anodised finish are important, choose 5356. However, ensure proper cleaning of the surfaces and use the correct shielding gas.

o If anodising is not required and you prefer a weld that flows more smoothly, 4043 is a good option. However, cleanliness is key to preventing porosity.
- **To avoid cracking:**
  o 4043 is better at preventing cracks, especially in crack-prone aluminium alloys.
  o 5356 is more likely to cause cracking issues, as it is less fluid and ductile than 4043.

CONCLUSION

In short:

✓ If you need **strength and durability** in **harsh environments, 5356** is the best choice.
✓ If you need **smooth flow and lower risk of cracking**, 4043 is the better filler material.

TIG WELDING ON ALUMINIUM 5086

PRACTICE 18. HORIZONTAL FILLET WELD—PA (1G)

- **Base material:** Two aluminium-magnesium plates, 5.91" × 1.57" × 0.118" (150 × 40 × 3 mm)
- **Electrodes:** Choose from Ø 3/32" (2.4mm) tungsten electrodes, either pure tungsten (WP green), zirconium alloyed (WZr3 brown), or rare earth alloyed (E3 purple).
- **Gas nozzle:** No. 7–8
- **Filler rod:** Ø 5/64"–5/32" (2–4mm) aluminium 5356
- **Tungsten extension:** 0.20"–0.24" (5–6mm) beyond the gas nozzle
- **Number of weld beads:** One bead deposited with a straight-line movement
- **Current intensity:** 60–90 amps
- **Gas flow rate:** 21–27 cubic feet per hour (CFH) (equivalent to 10–13L per minute), adjusted according to the nozzle diameter, using pure argon

WELDING PROCEDURE

1. **Preparation of the plates**

- **Degreasing:** Use a specialised aluminium degreaser or acetone to clean the plates. It is crucial to ensure the surface is completely free of oil and contaminants, as these can interfere with the welding process (Figure 2.35).

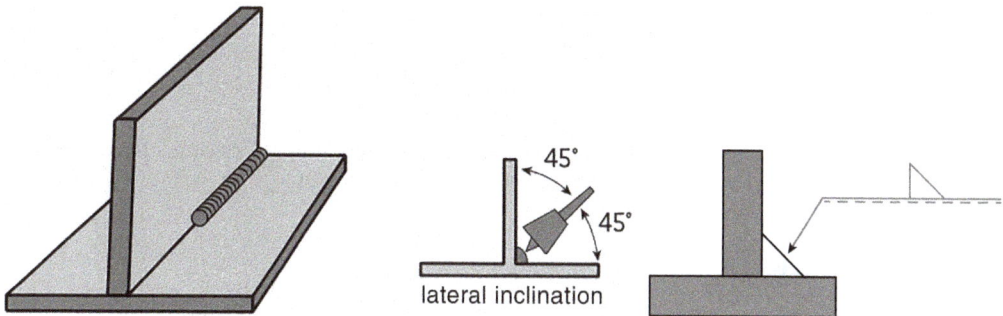

FIGURE 2.35 Fillet weld using the TIG process. The weld bead is visible at the joint, along with a diagram indicating the recommended 45° lateral electrode inclination.

2. Welding equipment setup

- **Pre-gas function:** Set the pre-gas flow to 1 second to purge the welding area of oxygen before starting the arc, reducing oxidation risks.
- **Post-gas function:** Set the post-gas flow to 10 seconds to protect the tungsten and the weld as they cool down, preventing oxidation.
- **Initial current ramp:** Program a current ramp from 150% to 100% over 3 seconds. This is essential to overcome aluminium's high thermal conductivity and to allow the weld pool to form in a controlled manner.
- **Final current ramp:** Set a 3-second current ramp from 100% to 30% to avoid craters and reduce the risk of cracks at the end of the weld.
- **4T mode:** Use 4T mode to gain better control over arc initiation and termination.
- **Pulsed arc:** Set the machine to 150 pulses per second (pps) with a wave balance of 75% in tungsten negative and 25% in tungsten positive. Start with 45 amps for the base current and 75 amps for the peak current, adjusting as necessary. These settings optimise surface cleaning while maintaining heat control during welding.

3. Welding process

- **Initial positioning:** Place the plates in a horizontal position, ensuring they are properly aligned and securely clamped.
- **Surface cleaning:** Just before welding, brush the joint area with a stainless steel wire brush to remove the aluminium oxide layer (alumina). This step is crucial, as alumina has a much higher melting point than aluminium (approximately 3,632°F vs. 1,220°F [2,000°C vs. 660°C]) and can cause weld defects if not removed properly.
- **Starting the weld:** Initiate the arc using the initial current ramp to quickly and smoothly form the weld pool.
- **Torch movement:** Instead of using the Walking the Cup technique, you will manoeuvre the torch in freehand mode with a controlled and meticulous approach:
  1. Pause the torch movement and wait for the weld pool to form.
  2. Introduce the filler rod at the edge of the weld pool to regulate the material input.
  3. Withdraw the filler rod and move the torch forward by a few millimetres.
  4. Repeat the process, noting that as the plates heat up, the weld pool will form more quickly. Adjust the current intensity if necessary to maintain control.
- **Temperature control:** If the weld pool forms too fast, stop and adjust the current to prevent overheating and plate distortion.

4. Completing the weld

- **Applying the final current ramp:** As you approach the end of the weld, use the final current ramp to gradually reduce arc power, preventing craters and defects at the weld termination.
- **Controlled cooling:** Allow the plates to cool gradually before handling them. This minimises distortion, as aluminium dissipates heat very quickly.

FINAL CONSIDERATIONS

This exercise presents a significant challenge due to aluminium's high thermal conductivity and the difficulty of controlling the weld pool. Precise temperature management and a steady torch movement are key to success. As you gain experience welding aluminium, you will realise that mastering these aspects is crucial for achieving high-quality weld beads.

TIG WELDING ON ALUMINIUM 5086

PRACTICE 18. BEVELLED PLATES IN "V" —HORIZONTAL POSITION (PA/1G)

- **Base material**: Two aluminium-magnesium plates, 5.91″ × 1.57″ × 0.118" ″ (150 × 40 × 3 mm)
- **Electrodes:** Choose from Ø 3/32″ (2.4mm) tungsten electrodes, either pure tungsten (WP green), zirconium alloyed (WZr3 brown), or rare earth alloyed (E3 purple).
- **Gas nozzle:** No. 7–8
- **Filler rod:** Ø 5/64″–5/32″ (2–4mm) aluminium 5356
- **Tungsten extension:** 0.20–0.24″ (5–6mm) beyond the gas nozzle
- **Number of weld beads:** One bead deposited with a straight-line movement
- **Current intensity:** 60–90 amps
- **Gas flow rate:** 21–27 cubic feet per hour (CFH) (equivalent to 10–13 litres per minute), adjusted according to the nozzle diameter, using pure argon

1. Preparation of the plates

Plate positioning:

- **Edge-to-edge plates:** If you decide to weld the plates without a root gap, it will not be necessary to protect the back side of the weld (Figure 2.36). The natural alumina layer (aluminium oxide) on the back will act as a barrier, preventing the weld pool from oxidising upon exposure to air.
- **Bevelled plates with a root gap:** If you choose to bevel the plates at 35° with a root gap (e.g. 0.10″/2.5mm), you must protect the back side using a ceramic or stainless steel backing strip. This prevents contamination from aluminium oxides, which would appear as black inclusions in the weld. Additionally, to keep the gap uniform during welding, you should tack-weld auxiliary plates every 2″ (5cm) along the back side. This will help counteract thermal expansion, which could close the gap and affect the weld quality.

Cleaning:

- **Brushing:** Just before welding, brush the joint area with a stainless steel wire brush to remove the alumina layer. This is essential for achieving a clean, defect-free weld.
- **Degreasing:** Use a specialised aluminium degreaser or acetone to clean both sides of the plates. Ensuring an oil- and contaminant-free surface is critical for high-quality welds.

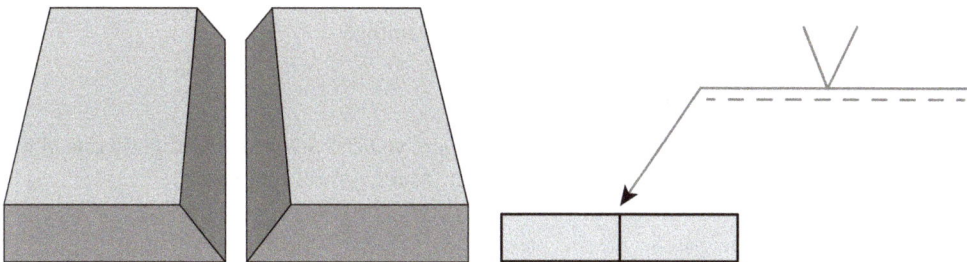

FIGURE 2.36   Preparation of a V-groove butt joint for welding. The gap between the pieces and the bevel angle on the edges are visible.

2. Welding equipment setup

- **Pre-gas and post-gas:** Use the same 1-second pre-gas and 10-second post-gas settings as in the previous practice.
- **Current ramps:** Configure:
  - o Initial current ramp: 150% → 100% over 3 seconds
  - o Final current ramp: 100% → 30% over 3 seconds
- **Four-stroke mode and pulsed arc:** Set the machine to four-stroke mode, with a pulsed arc at 150 pulses per second (pps), and a wave balance of 75% in tungsten negative and 25% in tungsten positive.

3. Welding process

Welding technique:

- **Starting the weld:** Initiate the arc with the initial current ramp to form the weld pool quickly and in a controlled manner.
- **Straight torch movement:**
  - o The Walking the Cup technique is not ideal for this practice due to aluminium's softness under heat.
  - o Instead, move the torch in a straight-line motion, keeping control:
    1. Pause when the weld pool forms.
    2. Introduce the filler rod at the edge of the weld pool to maintain control over deposition.
    3. Withdraw the filler rod and move the torch forward by a few millimetres.
    4. Repeat the process, adjusting the current intensity as needed to prevent excessive heat buildup.
- **Temperature control:**
  - o As you progress, heat buildup accelerates weld pool formation.
  - o If this becomes uncontrollable, pause and reduce the current to regain control.
- **Keep the plates elevated:**
  - o Avoid resting the joint on a workbench or support.
  - o Due to aluminium's high thermal conductivity, contact with a metal surface could drain heat away, preventing proper fusion at the weld edges.

4. Completing the weld

- **Final current ramp:** As you approach the end of the weld, reduce the current gradually using the final current ramp. This prevents crater formation and reduces the risk of cracks at the weld's endpoint.
- **Controlled cooling:** Allow the plates to cool gradually before removing auxiliary clamps, preventing distortions caused by rapid cooling.

5. Expected back-side weld appearance

If you successfully welded without bevels or a root gap, you should observe a slightly raised weld bead on the back side. This bead will look different from a typical root pass:

1. **Visible joint line:**
   - o The joint line will still be noticeable within the ridge of the penetration bead.

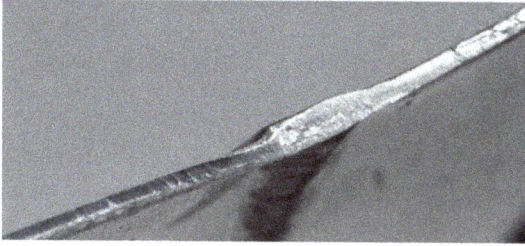

FIGURE 2.37   Close-up of a weld on a bevelled butt joint after the welding process. The metal fusion at the edge of the joint is visible.

2. **"Encased" appearance:**
   o   The weld may appear as if it is "trapped inside" a thin layer.
   o   This happens because aluminium (melting point ~1,220°F/660°C) melts, but the alumina layer on the back side (melting point ~3,632°F/2,000°C) does not.
   o   Imagine pouring water into a plastic bag—the water flows freely inside, but the bag itself remains intact, encasing the liquid (see Figure 2.37).

FINAL CONSIDERATIONS

This practice enhances your understanding of aluminium welding. While equipment setup is crucial, welding technique, temperature control, and material behaviour observation are key to achieving a high-quality weld.

Remember:

•   Persistence and attention to detail are essential when welding aluminium, given its high thermal conductivity and tendency to distort under heat.
•   If you want to practise on a qualification test coupon, here are the minimum dimensions (in millimetres and inches) according to UNE EN ISO 9606–2 for butt joints:

INSPECTION PROTOCOL FOR CSWIP WI FOR TIG WELDS IN ALUMINIUM 5086 SERIES

The inspection protocol for welder qualification in both butt joints and fillet welds using TIG welding in aluminium 5086 shares several common elements, as both types of joints require visual inspection, dimensional measurements, and in some cases, destructive and nondestructive testing. However, key differences exist due to the specific characteristics of each joint type (Figure 2.38).

Common aspects of both protocols

1. **Initial visual inspection**
   - o   Assessment of surface defects (cracks, porosity, oxides).
   - o   Uniformity of the weld bead.
   - o   Verification of cleanliness and absence of contaminants in the weld.
2. **Dimensional measurement**
   - o   Verification of base material thickness and weld bead dimensions.
   - o   Measurement of weld width and height.
3. **Destructive and non-destructive tests**
   - o   Bend tests to assess the ductility and mechanical strength of the weld.
   - o   Non-destructive testing, such as radiography or ultrasonic testing, to detect internal defects.

Key differences in the protocol

1.   *Joint configuration*

- • **Butt joint**
  - o   Focuses on penetration quality and root fusion. The weld must have complete penetration with an adequate root reinforcement, avoiding excessive reinforcement or undercuts.
  - o   **Misalignment:** Correct alignment is crucial to prevent issues such as lack of fusion or incomplete penetration.

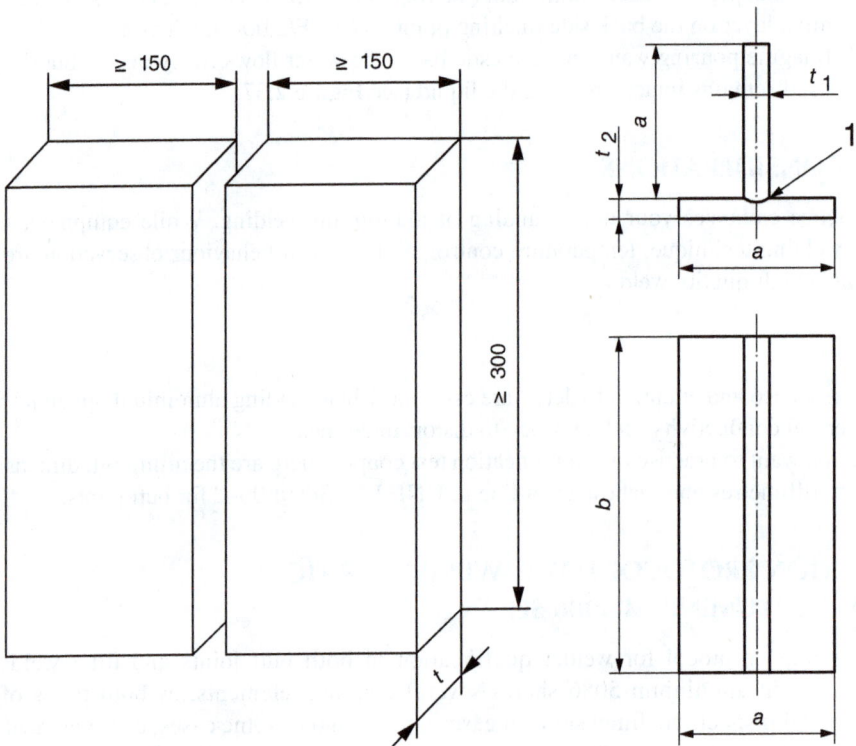

FIGURE 2.38   Dimensional diagram of a fillet and butt joint specimen. The minimum plate dimensions and thickness are specified for a welding test.

- **Fillet weld**
  - o Focuses on fusion quality at the vertex of the joint and symmetry of the weld bead on both sides.
  - o **Verification of the joint angle:** The 90° angle (or specified angle) is measured to ensure correct positioning during welding.
  - o **Reinforcement assessment:** The reinforcement must be consistent and symmetrical on both sides to ensure structural integrity.

2. Weld reinforcement evaluation

- **Butt joint**
  - o Reinforcement is assessed on both the root and face of the weld. Uniform reinforcement is required without undercuts or excessive material.
- **Fillet weld**
  - o The reinforcement is primarily evaluated at the joint vertex, ensuring no undercuts or excessive reinforcement. The symmetry of reinforcement on both sides is essential.

3. Symmetry and structural stability

- **Butt joint**
  - o Special attention is given to the weld bead symmetry to avoid stress concentrations.
  - o **Distortion:** The joint is checked for deformation caused by shrinkage during cooling.
- **Fillet weld**
  - o Ensures symmetry of the weld bead on both sides and overall structural stability.

4. Distortion evaluation

- **Butt joint**
  - o Checked to ensure that shrinkage and deformation do not affect alignment.
- **Fillet weld**
  - o Assessed to confirm that distortion does not alter the joint's angle or compromise its symmetry.

WELDER QUALIFICATION CRITERIA ACCORDING TO UNE EN ISO 9606–2

The following section includes key factors to consider when choosing the appropriate test for qualification:

- a) **Choosing the type of test**
  - • If most welding work involves fillet welds, the welder must be qualified with a fillet weld test.
  - • If most work involves butt welds, butt weld qualification automatically covers fillet welds.
  - • Knowing this distinction saves time and costs when selecting the test.
- b) **Filler material selection**
  - • If the qualification test is carried out using an AlMg filler wire (e.g., 5356, as used in the practice), the welder is also qualified to weld with AlSi alloys (e.g., 4043).
  - • However, if the test is conducted with AlSi (4043), the qualification does not cover AlMg alloys (5356).

c) **Thickness ranges for butt joints**
   - If the test coupon thickness is greater than 0.24″ (6mm), the welder is qualified for any thickness above 0.24″ (6mm).
   - If the test coupon is 0.24″ (6mm) or less, the qualification covers half to double the tested thickness.
     o (Example: A 0.12″ (3mm) coupon qualifies for thicknesses from 0.06″ (1.5mm) to 0.24″ (6mm)).
d) **Thickness ranges for fillet welds**
   - If the test coupon is 0.12″ (3mm) or greater, the welder is qualified for any thickness above 0.12″ (3mm).
   - If the test coupon is less than 0.12″ (3mm), the qualification covers thicknesses from the tested thickness up to 0.12″ (3mm).
     o (Example: A 0.06″ (1.5mm) coupon qualifies for thicknesses from 0.06″ (1.5mm) to 0.12″ (3mm)).
e) **Destructive and non-destructive testing**
   - Either a radiographic test, bend test, or fracture test must be performed.
   - Fracture tests can be replaced with a macroscopic examination of at least two sections.
f) **Certificate validity**
   - The welder qualification certificate is valid for a maximum of two years.

WHAT IF THE QUALIFICATION TEST IS "NOT ACCEPTABLE"?

As a professor once told me:

"I would rather have my wife angry at me than deal with aluminium welding issues . . . and you don't know how terrifying my wife can be when she's mad."

This insightful remark highlights that aluminium welding is challenging, and its complexity can easily lead to defects, causing a failed welder qualification test—even if the welder appears to be doing everything correctly. Fortunately, most of these issues can be prevented. Let's see how:

POROSITY PREVENTION: BEST PRACTICES

1. **Keep the material clean**
   o Clean the aluminium before welding to remove oil, grease, oxides, and dust.
   o Use acetone or a specific aluminium cleaner to eliminate contaminants that can cause gas entrapment (porosity).
2. **Check filler rods**
   o Ensure the filler wire is clean and dry. Dirty or moist rods introduce unwanted gases, leading to porosity.
3. **Optimise gas shielding**
   o Set the correct gas flow rate according to the nozzle size.
   o Use high-purity argon and check for gas leaks in the system.
4. **Inspect the gas nozzle**
   o Use the correct nozzle size and check regularly for blockages or damage.
5. **Maintain proper torch positioning**
   o Keep the torch at the correct angle and ensure proper gas coverage over the weld zone.

6. **Avoid draughts and air currents**
   o If welding in a windy area, use protective screens to prevent gas from being blown away.
7. **Control humidity**
   o Work in a dry environment and prevent condensation on workpieces, tools, and filler rods.
8. **Use proper welding technique**
   o Maintain a steady and controlled travel speed.
   o Avoid irregular movements that could trap air.
9. **Manage heat input**
   o Avoid overheating the material, as excessive heat can release contaminants, leading to porosity.
10. **Prevent cross-contamination**
    o If working with different metals, ensure there's no mixing of materials.
11. **Check the gas delivery system**
    o Inspect regulators, hoses, and fittings for leaks or mal.

THE MONK AND THE CUP OF TEA

A *young monk arrived at the house of a wise Tibetan master, seeking answers to his failures in meditation. Upon his arrival, the master offered him a cup of tea. The young monk accepted, but he could not help but start talking about his concerns.*

"Master, I have meditated for years, following the instructions to the letter, yet I keep failing. Why can't I attain the peace that others achieve?"

The master began pouring the tea. The monk's cup quickly filled, but the master continued pouring until the tea overflowed, spilling onto the table and the young monk's robes.

"Master! Can't you see the cup is full? It cannot hold any more tea!" exclaimed the monk.

The master smiled and said, "That is how you are now. Your mind is so full of preconceived ideas and fears of failure that there is no room left to learn anything new. Empty your cup, young monk, and the tea will be able to fill it once more. Failure is not the end, but an opportunity to learn and start again with humility and dedication."

TIG WELDING ON CAST IRON

PRACTICE 19. BEVELLED "V" GROOVE CAST IRON PLATES IN HORIZONTAL POSITION. PA(1G).

- **Base material:** Two pieces of cast iron, 6″ x 1.6″ x 0.2″.
- **Electrodes to be used:** Choose depending on availability between a 3/32″ diameter electrode alloyed with lanthanum (WLa 15 gold) or rare earths (E3 purple).
- **Nozzle size:** No. 7–8.
- **Filler rod:** 5/64–1/8″ ER Ni99 or ER NiFe-CI, designed for welding cast iron.
- **Electrode extension beyond nozzle:** 3/16–1/4″.
- **Number of passes:** One root pass + 1 hot pass.
- **Current intensity:** 70–110 amps.
- **Shielding gas flow rate:** 21–25 cubic feet per hour (CFH) (adjust flow rate according to nozzle diameter) of pure argon.
- **Additional precautions:** Preheat the workpiece to 300–480°F depending on thickness and ensure controlled cooling after welding to avoid cracking (Figure 2.39).

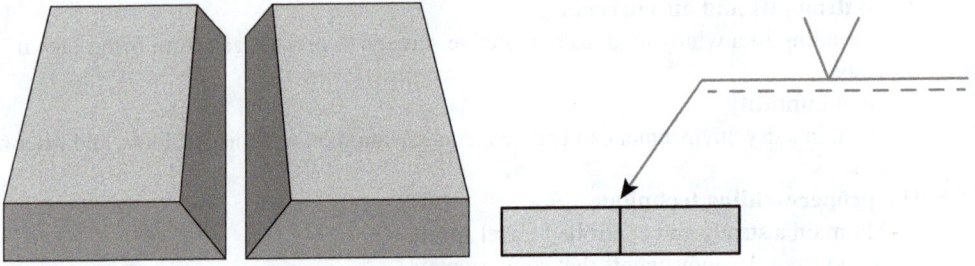

FIGURE 2.39   Preparation of a V-groove butt joint. Two metal pieces with bevels are shown, ready to be welded together.

DIFFERENCES BETWEEN CAST IRON AND CARBON STEEL

1. **Composition and structure:** Cast iron contains a higher carbon content (typically between 2–4%) compared to carbon steel, which results in a more brittle microcrystalline structure prone to cracking. This fundamental difference affects how the material should be prepared and welded.
2. **Weldability:** Cast iron is more challenging to weld than carbon steel due to its high rigidity and low ductility. When welded, it has a greater tendency to develop cracks, making it essential to control temperature and use filler materials specifically designed for this purpose.
3. **Heat treatments:** Unlike carbon steel, cast iron requires specific heat treatments before and after welding to minimise the risk of cracking and ensure weld integrity.

PREPARATION AND CLEANING PROCESS

1. **Surface preparation:**
   o **Preheating:** Before welding, preheat the cast iron pieces to a temperature between 300°F and 480°F. This helps reduce the temperature gradient between the weld zone and the rest of the material, minimising the risk of cracking during and after welding.
   o **Bevelling:** Bevel the edges of the pieces to 45° with a 3/32″ root gap to facilitate weld penetration. Bevelling also helps reduce stress concentration in the joint.
2. **Cleaning:**
   o **Removal of contaminants:** Thoroughly clean the pieces to remove any contaminants such as rust, grease, paint, or dust. Proper cleaning is crucial to prevent impurity inclusions in the weld, which could weaken the joint.
   o **Brushing:** Use a stainless steel wire brush to remove any surface residue. Although iron and steel are compatible, the presence of iron particles does not pose a significant contamination risk regarding base material compatibility. However, oxidation on iron particles (rust) could introduce impurities into the weld, leading to defects such as porosity or inclusions.

TACKING

1. **Applying tack welds:**
   o **Tack weld distribution:** Apply multiple tack welds around the joint to hold the pieces in place during the welding process. Tack welds should be placed at regular intervals to ensure an even distribution of stress.
   o **Expansion control:** During tacking, it is crucial to maintain constant preheating to prevent thermal expansion from closing or distorting the joint.

WELDING PROCESS

1. **Equipment Setup:**
   o **Electrode:** Use a lanthanum or rare earth alloyed electrode (3/32″ diameter). These tungsten types provide greater arc stability in the TIG process and are less prone to contamination.
   o Filler rod: Use an ER Ni99 or ER NiFe-CI filler rod. These alloys are designed for welding cast iron and have high resistance to crack formation.
   o **Shielding gas flow:** Adjust the shielding gas flow (pure argon) between 21–25 cubic feet per hour, adapting it to the nozzle diameter.
2. **Welding execution:**
   o **Welding technique:** In this case, the Walking the Cup technique is not recommended due to the rigid nature of cast iron. It is preferable to hold the torch freehand, manually controlling the movement to avoid excessive heat concentration in a single spot.
   o Bead deposition: Lay a root pass, ensuring proper penetration into the joint. Then, apply a hot pass to reinforce the weld and correct any defects left in the root pass.
   o **Temperature control:** Continuously monitor the temperature during welding to prevent overheating. Excessive heat can cause distortion or even cracks in the material.

POST-WELD HEAT TREATMENTS

1. **Controlled cooling:**
   o **Post-heating:** Once welding is complete, apply post-heating at approximately 300°F-400°F and maintain the pieces at this temperature for an extended period to relieve internal stresses. Avoid rapid cooling, as it can lead to cracking.
   o **Slow cooling:** Allow the pieces to cool gradually in an oven or cover them with insulating material (such as dry sand) to minimise crack formation during cooling.

STABILISATION TREATMENT

This heat treatment is applied after welding and before final cooling to improve dimensional stability and corrosion resistance in cast iron welds.

1. **What is stabilisation treatment?**
   o The stabilisation treatment consists of maintaining the welded piece at an elevated temperature for a prolonged period (typically between 930°F and 1110°F). This process allows internal residual stresses generated during welding to be evenly distributed and facilitates the homogeneous redistribution of carbides or other hardening phases within the metal structure.
2. **What is the purpose of the stabilisation treatment?**
   o **Reduction of internal stresses:** Applying this treatment relieves internal stresses that could cause distortions or cracks in the weld once the piece is in service.
   o Improved dimensional stability: Stabilisation helps minimise dimensional distortion, which may occur due to accumulated stresses during weld cooling.
   o **Enhanced corrosion resistance:** In some types of cast iron, stabilisation treatment helps reduce susceptibility to intergranular corrosion, a phenomenon that can weaken the material structure in corrosive environments.

3. **Application of the treatment:**
   o **Temperature and duration:** Maintain the piece at a temperature between 930°F and 1110°F for a period that may vary from 1 to 2 hours, depending on the size and thickness of the piece.
   o **Controlled cooling:** After the treatment, allow the piece to cool slowly to room temperature to prevent reintroduction of stresses.

This treatment is particularly useful when working with critical components where dimensional stability and corrosion resistance are essential. Applying this treatment ensures greater durability and reliability of welds on cast iron.

COMMON DEFECTS IN TIG (GTAW) WELDING

1. Porosity

What is it?

Porosity in TIG welding appears as small cavities or bubbles trapped inside or on the surface of the weld bead. These bubbles form when gases become trapped in the molten metal before it solidifies.

Why does it occur?

- Contamination on the base material or filler rod (oxide, grease, oil, or moisture).
- Insufficient or incorrect shielding gas flow (argon, helium, or mixture).
- Air currents displacing the shielding gas.
- Excessive arc length, exposing the weld pool to the atmosphere.

How to prevent it?

- Thoroughly clean the base material and filler rod before welding.
- Adjust the shielding gas flow according to the nozzle size and environmental conditions (typically between 17–32 cubic feet per hour—CFH).
- Protect the welding area from air currents and use welding screens if necessary.
- Maintain a short and consistent arc length to ensure effective gas coverage.

2. Tungsten contamination

What is it?

Contamination occurs when the tungsten electrode comes into contact with the weld pool or filler rod, leaving inclusions in the weld bead.

Why does it occur?

- Poor technique in handling the electrode.
- Use of an improperly sharpened or worn tungsten electrode.
- Excessive current, causing the tungsten tip to melt.

How to prevent it?

- Maintain a constant electrode distance and avoid touching the weld pool or filler rod.
- Sharpen the tungsten in a parallel direction with an appropriate angle for the type of current (DC or AC).
- Use the correct current for the electrode diameter.

3. Lack of fusion

What is it?

Lack of fusion occurs when the filler metal does not completely fuse with the base material or previous weld passes, leading to weak joints.

Why does it occur?

- Insufficient current, generating inadequate heat.
- Excessive welding speed or excessive filler metal application.
- Incorrect torch positioning or angle.

How to prevent it?

- Adjust the current according to the base material thickness.
- Reduce the welding speed to allow enough time for the arc to fuse the materials properly.
- Maintain an appropriate torch angle (typically between 10° and 15° in the direction of travel).

4. Weld bead oxidation

What is it?

Weld bead oxidation occurs when the molten metal or HAZ is not fully shielded by the inert gas, resulting in a dark or defective weld bead.

Why does it occur?

- Insufficient or interrupted shielding gas flow.
- Use of small nozzles that do not provide adequate coverage.
- Lack of internal purging in materials such as stainless steel or nickel alloys.

How to prevent it?

- Ensure the gas flow is steady and adequate.
- Use a larger nozzle or a gas diffuser to improve coverage.
- Apply internal purging with an inert gas when welding materials prone to oxidation.

5. Oxide inclusions

What is it?

Oxide inclusions are trapped particles within the weld bead due to poor cleaning or incorrect welding technique.

Why does it occur?

- Base material or filler rod with oxide or contaminants.
- Poor technique when feeding the filler rod, allowing it to touch oxidised areas.
- Lack of gas coverage at the edges of the weld pool.

How to prevent it?

- Thoroughly clean the surfaces and filler rod before welding.
- Feed the filler rod only within the area protected by the inert gas.
- Maintain uniform gas coverage over the weld pool.

PRACTICAL SOLUTIONS TO PREVENT AND CORRECT DEFECTS

1. Adjusting welding parameters

- **Current:** Select the correct current according to the base material thickness and type of weld.
- **Gas flow:** Adjust the flow rate to ensure efficient shielding of the weld pool.
- **Arc length:** Maintain an appropriate distance between the tungsten and the base material (usually 0.08–0.12").

2. Welding technique

- **Torch angle:** Keep the torch inclined in the direction of travel to optimise gas coverage and facilitate filler rod feeding.
- **Torch movement:** Use a steady and consistent movement to avoid fluctuations in the weld pool.
- **Cleaning:** Remove any oxide or contaminants before welding to prevent defects in the weld bead.

3. Environmental control

- Use screens to protect the welding area from air currents.
- Work in a clean and dry environment to minimise contamination and oxidation.

4. Equipment maintenance

- Regularly check the nozzle, diffuser, and gas hoses for leaks or wear.
- Ensure the tungsten electrode is sharpened and clean.

CONCLUSION

TIG welding requires precision and attention to detail to achieve high-quality results. Identifying and correcting common defects, as well as adjusting parameters and improving welding techniques, are essential steps to ensure durable and aesthetically pleasing welds. With practice and a consistent focus on cleanliness and control, you can master this process and produce outstanding welds.

Do you want more information? Here you can find more information about TIG welding:
Facultad de Soldadura:
www.youtube.com/watch?v=FFm3BciuTKA
www.youtube.com/watch?v=2H9XvbmDhDE

3 MIG/MAG/Cored Wire Welding

WISDOM *"THE IMPORTANCE OF CALMNESS"*

The master insisted time and again on serenity.

- Let your mind settle, calm down, and find peace.
- But why do you consider tranquillity so important?
- Come with me, the master requested.

He led his disciple to a pond and began stirring the water with a stick. Then, he asked:

- Can you see your face in the water?
- How could I? The water is disturbed! It's impossible, the disciple protested, thinking the master was mocking him.
- In the same way, as long as you are agitated, you will not be able to see the face of your inner self. In the comforting stillness of the mind, when the clamour of thoughts is silenced, the voice of the inner being will emerge.

Yoga. El silencio es mi alimento—*Vicente Moreno*

It is not about silencing the mind—that is impossible. It is about not identifying with what it shows. Only in calmness can we think clearly.

INTRODUCTION TO MIG/MAG WELDING

In this chapter, we will cover MIG/MAG welding, also known as **"semi-automatic welding"** or **"wire welding".** This process is more complex compared to shielded metal arc welding (SMAW) and TIG welding, making it necessary to understand its fundamentals, proper equipment parameterisation, and maintenance.

Historical origins of MIG/MAG welding

By the 1950s, SMAW and TIG welding were already well-established, but they presented certain limitations:

- **SMAW:** It was unsuitable for mass production due to the time spent replacing consumed electrodes and removing slag. Additionally, it was difficult to automate.
- **TIG welding:** While effective for reactive materials, it was not cost-effective for thicknesses above 4–6mm.

The solution to these limitations was the invention of **Metal Inert Gas (MIG) welding,** a process that replaced the tungsten electrode with a continuously fed wire from a spool. This process not only overcame the limitations of SMAW but also provided higher-quality weld beads, especially for materials sensitive to contamination. Additionally, it was much faster and more cost-effective for welding thicker sections compared to TIG.

DOI: 10.1201/9781003422488-4

However, for welding steel, the cost of inert gases such as argon and helium was high. This led to the development of **Metal Active Gas (MAG) welding,** which used more affordable gases such as pure CO_2 or CO_2 mixed with argon.

FUNDAMENTALS OF THE MIG/MAG PROCESS

Required equipment:

- **Power source:** Equipped with a transformer-rectifier to operate with direct current. Unlike SMAW and TIG welding equipment, the power source here has a constant voltage characteristic.
- **Wire feed system:** Includes a torch, wire spool, and a cooling system for the equipment and torch, if necessary.
- **Gas supply:** Protects the weld pool.
- **Electromagnetic valves and regulators:** Synchronise cooling (if applicable), gas flow, and electrical current.

COMPONENTS OF THE MIG/MAG EQUIPMENT

MIG/MAG welding is **semi-automatic**—the wire is fed automatically, but the welder manually controls the process. The key components are (see Figure 3.1):

- **Wire feeding system:** The wire is automatically fed through rollers driven by a motor, synchronised with the torch trigger. These rollers have different grooves depending on the type of wire used (V-groove for solid wires, U-groove for soft wires). It is crucial to keep this system clean to prevent contamination and welding defects.
- **Nozzle, contact tip, and diffuser:**
  o   The **nozzle** directs the shielding gas to the weld pool.
  o   The **contact tip** transmits electrical current to the wire.
  o   The **diffuser** connects the contact tip to the torch, distributing the gas.
- **Wire spool holder ("reel stand"):** Holds the wire spool and includes a brake to stop the spool's inertia when welding stops.
- **Hose:** Protects the wire conduit, shielding gas, power cable, and cooling circuit (if applicable). Careful handling is necessary to avoid damage.
- **Trigger and torch:** The trigger starts or stops the welding process, while the torch guides the wire towards the weld seam.

KEY CONCEPTS IN MIG/MAG WELDING

- **Current (amperage):** The number of electrons flowing through a conductor in a given time, measured in **amperes (A).**
- **Voltage (potential difference):** The **"push"** that moves electrons towards the positive pole, measured in **volts (V).**
- **Electrical resistance:** The opposition to electron flow in a conductor.
- **Joule effect:** The heat generated when an electric current passes through a conductor, proportional to resistance, time, and current.

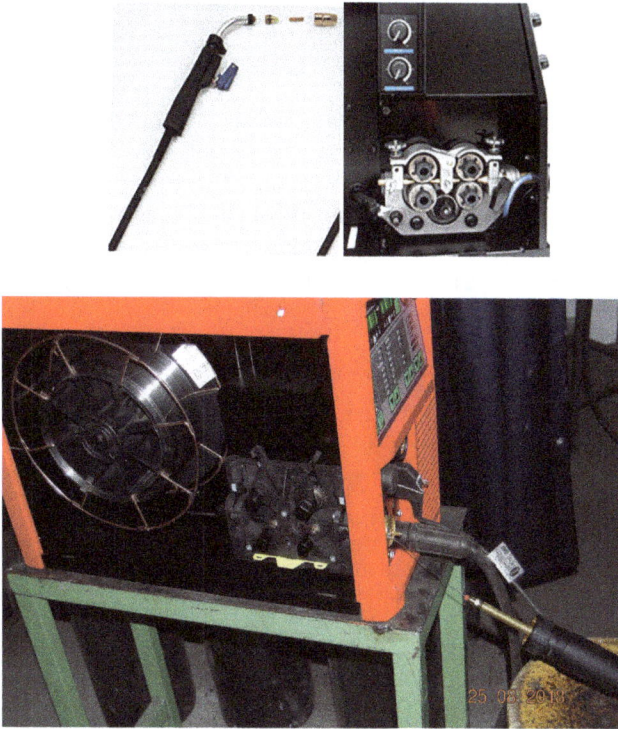

FIGURE 3.1 Series of images of a MIG/MAG welding machine: the wire feeding mechanism, a disassembled torch with its components, and an open welder showing the wire spool and feeding system.

MIG/MAG welding differs in its power source type:

- **Constant current:** Used in SMAW and TIG welding, where the priority is to maintain a steady current.
- **Constant voltage:** Used in MIG/MAG welding, where voltage remains constant regardless of the torch-to-workpiece distance.

TRANSFER MODES

Depending on current, wire feed speed, gas type, and wire diameter, the transfer mode varies:

- **Short-circuit transfer:** Used for thin or medium thicknesses, where the wire touches the workpiece, creating an arc that melts the tip into the weld pool.
- **Globular transfer:** Similar to short-circuit but with larger droplets.
- **Spray transfer:** At higher voltages, the wire is atomised into fine droplets that fuse with the workpiece, providing deeper penetration.
- **Pulsed arc transfer:** A method that alternates between spray and short-circuit transfer, offering better control and weld bead appearance.

ARC LENGTH

Arc length is the distance between the wire tip and the weld pool on the base material. This parameter directly affects weld bead shape and penetration:

- **Increased arc length:** A longer arc spreads the heat over a larger area, resulting in a wider but shallower weld bead.
- **Decreased arc length:** A shorter arc concentrates the heat on a smaller area, leading to deeper but narrower penetration.

HOW TO ADJUST IT

Some machines allow arc length adjustments via the control panel.

OHM'S LAW APPLIED TO **MIG/MAG** WELDING

Ohm's Law is a fundamental relationship in electricity, stating that:

$$Voltage(V) = Current(I) \times Resistance(R)$$

Applied to MIG/MAG welding:

- **Increasing the stick-out length (wire extension):** If you extend the distance between the torch and the workpiece, the circuit resistance increases (see Figure 3.2). According

FIGURE 3.2 Comparison of two types of arc height in MIG/MAG welding. On the left, deep penetration welding; on the right, superficial fusion welding.

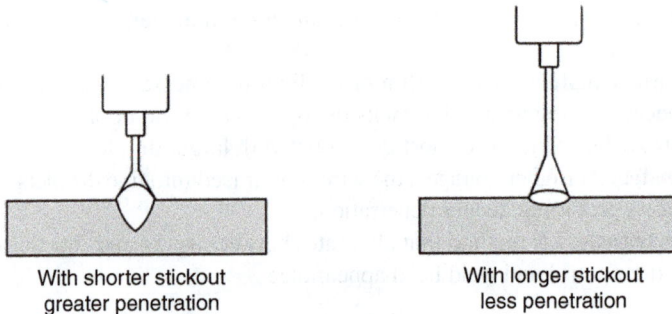

With shorter stickout
greater penetration

With longer stickout
less penetration

FIGURE 3.3 Comparative illustration of penetration in MIG/MAG welding. On the left, deeper penetration with a short arc; on the right, shallower penetration with a long arc.

to Ohm's Law, when resistance rises, current decreases to maintain a constant voltage. This reduces weld penetration, as less energy reaches the weld pool.

- **Reducing the stick-out length:** Shortening the distance lowers resistance, increasing current and thereby enhancing weld penetration (see Figure 3.3).

How to apply this concept: If you need deeper penetration, keep the wire as short as possible without allowing the contact tip to touch the workpiece. If less penetration is required, extend the stick-out distance.

PUSHING VS PULLING THE TORCH

- **Pushing the torch:** When welding, if you push the torch forward in the direction of travel:
  - **Weld bead shape:** Produces a wider, shallower bead, as the shielding gas covers the weld pool more effectively but spreads the heat over a larger area.
  - **Applications:** Ideal for welds requiring a cleaner surface finish and less penetration, such as thin sheet metal welding.
- **Pulling the torch:** When directing the torch backward, opposite to the travel direction:
  - **Weld bead shape:** Produces a narrower and deeper bead, as the heat concentrates on a smaller area, increasing penetration.
  - **Applications:** Best suited for welds requiring greater penetration, such as thicker materials.

How to apply these concepts: Depending on material thickness and penetration requirements, push the torch for a smoother, flatter bead or pull it for deeper penetration.

EFFECT OF TRAVEL SPEED

Definition of travel speed: Travel speed refers to how fast the welder moves the welding torch along the joint. This parameter significantly affects weld quality, including penetration, bead shape, and potential defects such as porosity or undercutting.

Practical examples:

- **High travel speed:** If the welder moves the torch too quickly, the arc does not have enough time to properly heat the base material. This can lead to insufficient penetration and a narrow, cold bead. This is particularly problematic in thick materials, where lack of fusion can compromise weld strength.
- **Low travel speed:** If the speed is too slow, the base material may overheat, leading to undercutting (erosion of the joint edges), excessive penetration, and even material distortion. Additionally, a slow travel speed increases the risk of porosity and inclusions due to prolonged exposure of the weld pool to atmospheric gases.

Recommendations:

- Adjust travel speed based on material thickness and equipment settings. Practise on test pieces to find the optimal speed that produces a uniform bead with good penetration without defects.
- Pay attention to arc sound and weld pool behaviour. A steady, uniform sound usually indicates an appropriate travel speed.

INDUCTANCE (ALSO CALLED "DYNAMICS")

Inductance is a parameter that controls how quickly the current responds to abrupt changes in demand during MIG/MAG welding. It directly affects bead fluidity and spatter levels:

- **High inductance:** Slows down the current response, resulting in a smoother weld bead with less spatter. However, it may slightly reduce penetration.
- **Low inductance:** Makes the current react more quickly and aggressively, leading to a rougher bead with more spatter but increased penetration.

How to apply this concept: If you seek a cleaner and smoother finish, increase inductance. If penetration is more important and spatter is not a concern, reduce inductance.

STICK-OUT LENGTH (WIRE EXTENSION FROM THE CONTACT TIP)

In MIG/MAG welding, maintaining a consistent wire extension from the contact tip to the weld pool is essential. If this length varies too much, problems may arise, such as the wire arriving "cold" at the workpiece, preventing proper fusion and causing it to bounce instead of forming a weld pool.

Concept: As wire extension increases, so does its electrical resistance. This reduces current flow, potentially causing welding defects. Additionally, a longer wire extension reduces the effectiveness of the shielding gas, which can also impact weld quality.

Recommended practice: Keep a consistent wire extension between **5mm and 15mm,** depending on the welding position and material type (specified in each exercise).

SHIELDING GASES

In MIG/MAG welding with solid wires, it is essential to use a shielding gas that protects the arc and weld pool from oxygen and nitrogen in the air. These gases can cause defects like porosity. Air currents and improper gas flow rates (too high or too low) can also create issues.

Types of gases and their applications:

- **Argon:** Provides arc stability and effective shielding due to its high density (1.4 times heavier than air). It produces minimal spatter and is less sensitive to air currents. Weld beads tend to be smooth and narrow.
- **Helium:** Offers high thermal conductivity, promoting deep penetration and wider beads. However, its arc stability is lower, and since helium is less dense than argon, it is more affected by air currents. It requires higher flow rates and is more expensive.
- **Argon-Helium mixtures (70–30% or 50–50%):** Combine the advantages of both gases. Ideal for welding aluminium over 8mm thick, reducing the risk of lack of fusion at the joint edges.
- **CO_2:** More cost-effective and primarily used for welding carbon steels with deoxidised wires. While CO_2 introduces some oxygen, which can be detrimental in most welds, its low cost and excellent penetration make it useful where extremely high-quality welds are not required.
- **Argon-CO_2 mixtures:** Commonly used for welding carbon steels, with typical compositions such as Ar 80%-CO_2 15% (also 82–18% or 85–20%) to enhance arc stability while maintaining weld quality.
- **Argon-CO_2-O_2 mixtures:** Used for welding carbon and stainless steels, improving arc stability, bead fluidity, and reducing spatter. Oxygen (usually 1–5%) helps stabilise the

arc and improves bead fluidity by lowering the weld pool's surface tension, resulting in a smoother finish.

Applications:

- **Argon:** Aluminium and magnesium.
- **Argon + CO_2 (1–5%):** Stainless steels, alloyed steels, and copper.
- **Argon + CO_2 (up to 20%):** Carbon steels.
- **CO_2:** Carbon steels with deoxidised wires.
- **Helium:** Aluminium, magnesium, and copper.
- **Helium + Argon (20–80% or 50–50%):** Aluminium, magnesium, and copper.

FILLER WIRES

Solid wires used in MIG/MAG welding come in various diameters: 0.6mm, 0.8mm, 1mm, 1.2mm, 1.4mm, 1.6mm, and 2.4mm.

Types of wire:

- **Carbon steel wires:** Usually coated with copper to improve conductivity, enhance corrosion resistance, and reduce friction during wire feeding. However, some wires are now available **without copper coating,** treated with a special passivation layer to prevent corrosion, eliminating the risk of copper contamination in the weld.
- **Flux-cored wires:**
  - **Rutile type:** Suitable for short-circuit or spray transfer. The slag is easy to remove and cools quickly, making these wires suitable for all positions.
  - **Basic type:** Produces stronger, higher-quality welds. Although the slag is harder to remove, these wires are also suitable for all welding positions.
- **Self-shielded flux-cored wires:** Contain elements that generate shielding gases during welding, eliminating the need for an external gas supply. Ideal for outdoor applications or windy conditions. However, they produce more slag and a less refined finish than solid wires.
- **Metal-cored wires:** Offer **high deposition rates** and **good penetration.** These wires contain **metal powder** inside a hollow core, improving heat transfer to the base material and making more energy available for fusion. Like flux-cored wires, they require external shielding gas.

SELECTION OF FILLER WIRE DIAMETER

IMPORTANCE OF DIAMETER

The diameter of the filler wire is a key factor in MIG/MAG welding, as it determines the amount of material deposited in the weld bead and how the arc behaves during the process. A smaller wire diameter is easier to control, making it ideal for detailed work or welding thin sheets. Conversely, a larger wire diameter allows for a higher deposition rate, which is advantageous for welding thicker materials or when high productivity is required.

PRACTICAL EXAMPLES

- **0.6mm or 0.8mm wires:** Ideal for welding thin sheets up to 3mm thick, where good arc control is necessary to avoid excessive penetration or deformation. This diameter is commonly used in automotive applications or sheet metal work.

- **1mm or 1.2mm wires:** The most commonly used sizes in general industrial applications, such as welding steel structures and medium-thickness pipes (3–10mm). They offer a balance between control and deposition speed, making them versatile for a wide range of tasks.
- **1.6mm or larger wires:** Suitable for welding thick materials, generally over 10mm, where a high deposition rate is needed. Used in the fabrication of large metal structures, such as in the shipbuilding and heavy construction industries, where speed and deep penetration are essential.

WELDING ASSISTANCE SYSTEMS

Some of the main assistance systems in MIG/MAG welding include:

- **Two-stroke and four-stroke modes:**
  - o **Two-stroke:** Pressing the trigger activates gas, current, and wire feeding. Releasing it stops the entire process.
  - o **Four-stroke:** Pressing the trigger only activates the gas, allowing the area to be purged before welding begins. Releasing it activates the current and wire feed, allowing continuous welding without holding the trigger. Pressing the trigger again stops the current and wire feed but allows the gas to continue flowing, protecting the weld pool as it cools.
- **Pre-flow and post-flow of shielding gas:** These allow gas flow to be programmed before and after welding starts or stops.
  - o **Pre-flow:** Purges the starting area to eliminate contaminants.
  - o **Post-flow:** Protects the weld pool as it cools, reducing crater formation.
- **Slow start and crater fill (burn-back):** Some machines allow the wire and current to start at a slower rate for easier bead overlapping. The crater fill function gradually reduces current and wire feed at the end of the weld, filling craters and improving bead quality.

MIG/MAG EQUIPMENT MAINTENANCE

Importance of maintenance

Regular maintenance of MIG/MAG equipment is essential to ensure weld quality, extend equipment lifespan, and prevent unexpected breakdowns. Well-maintained equipment ensures a steady flow of wire and gas, efficient current transfer, and adequate weld pool protection.

Basic maintenance steps

- **Cleaning the torch and nozzle:** Remove spatter and residues after each welding session to ensure proper gas flow and smooth wire feeding.
- **Checking the feed rollers:** Rollers must be free of grease, dirt, and wear. Dirty or worn rollers can cause wire feeding issues, leading to defects in the weld.
- **Inspecting and cleaning the hose:** The hose connecting the torch to the power source should be in good condition, without cracks or obstructions. The inner wire conduit should be cleaned or replaced periodically.
- **Checking electrical and gas connections:** Ensure all connections are tight and corrosion-free. Gas leaks or loose electrical connections can reduce efficiency and cause welding defects.

Practical tip

- Implement a regular maintenance schedule, checking and cleaning key components weekly or after intensive use. This not only improves weld quality but also reduces the likelihood of costly repairs.

WELDING ERGONOMICS

Posture and ergonomics

Maintaining a proper posture during welding is crucial to prevent fatigue and reduce the risk of long-term injuries. Good ergonomics also help maintain precise torch control, essential for high-quality weld beads.

Practical tips

- **Body positioning:** Keep your body in a balanced and stable position. If possible, rest your arms on a firm surface to prevent unsteady movements that could affect welding accuracy.
- **Work height:** Adjust the workbench height so that you can weld comfortably without hunching over. For extended welding sessions in awkward positions, consider using supports or pads to reduce back and leg strain.
- **Use of support tools:** Use magnetic holders or clamps to secure workpieces. This allows you to focus on torch control rather than the stability of the pieces.

WELDING SAFETY

MIG/MAG welding involves risks such as ultraviolet radiation exposure, fume inhalation, and burn hazards. Following strict safety practices is essential to protect your health.

Safety recommendations

- **Personal protective equipment (PPE):** Always wear welding gloves, an auto-darkening helmet, flame-resistant clothing, and safety boots. Ensure clothing fully covers the skin to prevent burns from radiation or spatter.
- **Proper ventilation:** Work in a well-ventilated area to avoid the accumulation of welding fumes. If working in an enclosed space, use fume extraction systems or appropriate respirators.
- **Safe equipment handling:** Always disconnect the power source before performing any maintenance on the torch or machine. Regularly check cables and connections for damage to prevent short circuits or electric shocks.

ENVIRONMENTAL FACTORS AFFECTING WELDING

The influence of surroundings on welding

Welding does not take place in a perfect environment—it is affected by various environmental conditions that can impact weld bead quality and efficiency. Understanding these factors and how to manage them is key to achieving consistent, high-quality results.

1. **Temperature**
   - **Impact on welding:** Ambient temperature variations can affect the base material's ability to retain heat.

o   Cold environments: The base material may cool too quickly, leading to excessive contraction, residual stress, and cracking.
o   Hot environments: The material may overheat, making the weld pool difficult to control.
- **Mitigation measures:**
o   Preheat workpieces in cold conditions to slow down cooling and reduce cracking risks.
o   Monitor heat input in hot conditions to prevent overheating.

2. **Humidity**
- **Impact on welding:** Moisture can introduce hydrogen into the weld pool, increasing the risk of porosity. It can also cause oxidation on the base material and consumables.
- **Mitigation measures:**
o   Store consumables in dry conditions.
o   Clean the base material before welding to remove moisture or oxidation.

3. **Air currents**
- **Impact on welding:** Air currents can displace shielding gas, exposing the weld pool to oxygen and causing porosity.
- **Mitigation Measures:**
o   Use welding screens or curtains to block airflow.
o   Adjust shielding gas flow rates based on environmental conditions.
o   Consider self-shielded flux-cored wires for outdoor welding.

General mitigation measures

1. **Wind screens:** An effective solution to protect the welding area from air currents is to use welding screens or curtains. These physical barriers reduce the exposure of the weld pool to the air, ensuring the shielding gas remains in place and improving weld quality.
2. **Preheating workpieces:** In cold environments or when working with materials prone to cracking, preheating can be a key technique. This helps maintain a consistent temperature during welding, reducing thermal stresses and minimising the likelihood of cracks in the weld bead.
3. **Shielding gas flow control:** Adjusting the shielding gas flow rate according to environmental conditions is essential.
o   In windy conditions, increasing the gas flow ensures that the weld pool remains adequately protected.
o   However, excessive gas flow should be avoided, as it can cause turbulence and draw air into the weld pool, leading to defects.
4. **Working in controlled environments:** Whenever possible, perform welding in a controlled environment. This includes not only temperature and humidity regulation but also airflow control. Working indoors or in a closed workshop provides better control over environmental factors and facilitates the production of high-quality welds.

CONCLUSION

Although often overlooked, environmental factors have a significant impact on the welding process. Whether temperature, humidity, or air currents, each of these elements can alter weld bead quality if not properly managed.

By implementing effective mitigation measures and understanding how each factor affects welding, you can adapt your technique to environmental conditions, ensuring consistent, high-quality results in every job.

FINAL CONSIDERATIONS

MIG/MAG welding is versatile and highly effective, but its complexity requires a solid understanding of equipment and processes.

Although synergic machines exist that automate some parameters, it is highly recommended to learn with conventional machines to develop a complete understanding of the welding process.

THE HIDDEN TREASURE

High up on a mountain lived an old guardian of a legendary treasure. Many adventurers came seeking it, but none could find it. Impatiently, they dug in all the wrong places, and after a few days, they left frustrated, claiming the treasure did not exist.

One day, a young apprentice climbed the mountain and approached the old man.

"Master, I seek the treasure, but I don't know where to start," he humbly said.

The old man smiled and handed him a shovel, pointing to a random spot.

"Dig here, but remember: the treasure is not just in what you find, but in what you learn while searching for it."

Day after day, the young man dug tirelessly, facing roots, rocks, and his own exhaustion. He learned to listen to the earth, understand the strength of his hands, and respect the land. Weeks passed, and though he did not find gold coins, he realised something inside him had changed—his patience, focus, and inner calm had grown.

One day, the old man returned and saw the young man working peacefully.

"You have found the treasure," he said. "It is within you, in every stroke of the shovel, in every challenge you overcame. Now you are wiser and stronger than when you arrived. Remember, the true treasure is not gold, but what you build while searching for it."

VERY IMPORTANT: BEST PRACTICES IN THE WELDING WORKSHOP

- **Always wear safety glasses:** The copper coating on welding wire, especially in carbon steel, can unexpectedly splinter off when the weld cools. Protect your eyes to avoid injuries.
- **Detecting porosity:** If you notice porosity forming in the weld pool (those tiny "black spots" in the weld bead), stop immediately. Use a grinder to completely remove the contaminated area before continuing. If you don't remove the porosity, the entire weld is likely to be defective.
- **Use the correct contact tip:** Choose the appropriate tip for the wire diameter.
- **Maintain the torch nozzle:** Regularly clean the nozzle and internal parts of the torch to remove spatter that can contaminate the weld and block the flow of shielding gas. Consider using an anti-spatter spray to slow down residue build-up.
- **Take care of the welding hose:** Keep the hose as straight as possible to prevent wire feed issues. Avoid dropping objects on the hose, stepping on it, or bending it excessively, as this can damage the liner that guides the wire.
- **Correct pressure on the drive rollers:** Always use the minimum necessary pressure on the drive rollers, adjusted according to the welding position. This helps prevent wire crushing or feeding problems.
- **Torch visibility and handling:** If you've practised with SMAW, you may find MIG/MAG welding easier. However, keep in mind that the MIG/MAG arc is not as bright, which can reduce visibility. Additionally, the nozzle may obstruct your view of the work area. Get used to looking from the side of the torch while keeping the wire tip in sight.

- **Maintain a constant wire stick-out:** Try to keep the wire length as constant as possible from the contact tip to the weld seam.
  - o If you move the torch away, voltage decreases.
  - o If you move the torch closer, voltage increases.
  - o These changes affect weld bead quality.
- **Use tools to handle hot parts:** Always use pliers or tongs to handle hot pieces—never your hands, even if you are wearing welding gloves. Metal pieces can be extremely hot and cause severe burns.
- **Ask your instructor:** If you have any doubts about a task, especially when using grinders, saws, or abrasive wheels, don't hesitate to ask your instructor or supervisor. These machines require careful and respectful use with all necessary safety precautions.
- **Eye protection:** Never look directly at the welding arc without the proper welding helmet. Direct exposure can cause serious eye damage.
- **Personal Protective Equipment (PPE):**
  - o Never weld without gloves or with short sleeves.
  - o Always wear the necessary protection: apron, sleeves, spats, etc.—it's for your safety!
  - o At the end of the day, return your PPE to its designated place.
  - o Also, consider wearing an appropriate respirator to protect against welding fumes.
- **Material care and cleanliness:**
  - o Be mindful of your tools and materials.
  - o Keep your workspace clean and organised after each session.
  - o A clean and orderly environment is essential for safe and efficient work.

MAG WELDING WITH CARBON STEEL

PRACTICE 1. FIRST WELD BEADS IN HORIZONTAL POSITION PA (1G)

- **Base material:** Carbon steel plate 4″ x 4″ x 1/8″
- **Wire diameter and designation:** 0.030″ (0.8mm) ER 70S-6 (AWS A5.18–05) G 46 3 M 2Mo (EN ISO 14341-A:2008)
- **Number of weld beads:** Nine, using a straight motion with the torch directed forward (pushing technique)
- **Welding voltage:** 15 to 18 volts
- **Wire feed speed:** 10 to 23 feet per minute
- **Stick-out length:** 3/8″ to 5/8″ (measured from the contact tip to the weld bead)
- **Gas flow rate:** 21 to 32 cubic feet per hour (CFH) of argon (85%)/CO_2 (15%) (approximately 1 CFH per 0.04″ of nozzle inner diameter)
- **Auxiliary tools:** Ruler, scriber, measuring tape, file, grinder, centre punch, and hammer (Figure 3.4)

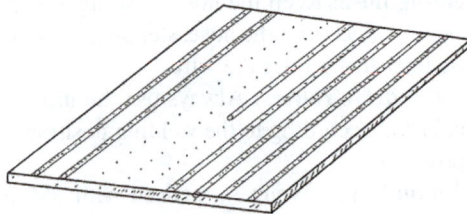

FIGURE 3.4   Metal plate with multiple parallel weld beads, used for straight-line welding practice. The uniform arrangement of the beads suggests an exercise in arc control and material deposition.

MATERIAL AND TOOL PREPARATION

1. **Cutting and preparing the workpiece**
   - **Measuring and marking:** Start by measuring the carbon steel flat bar with a ruler or measuring tape to ensure the final dimensions are 4″ x 4″. Mark the cutting lines with a scriber and highlight them using a centre punch to clearly define where you will cut.
   - **Cutting the workpiece:** Use a grinder fitted with a suitable cutting disc to cut the flat bar following the marked lines.
   - **Surface cleaning:** After cutting, remove any rust, grease, or coolant residues using a grinder with a flap disc or grinding wheel. Be careful not to remove too much material thickness during this process.
   - **Deburring:** Once the surface is clean, remove burrs from the cut edges and round off the corners of the plate using a file or grinder.
   - **Marking guide lines:** Using a scriber, mark a line every 3/8″ on two opposite edges of the square. Using a set square, draw straight lines connecting these marks—these will be your guidelines for welding. Add centre punch marks along these lines to improve visibility during welding.

2. **Equipment setup**
   - **Gas flow adjustment:** Set the shielding gas flow rate to 21–32 cubic feet per hour of the argon-CO_2 mixture, considering the inner diameter of the nozzle.
   - **Operating mode:** Choose whether to weld in two-stroke or four-stroke mode, depending on your preference and comfort. If the equipment is synergic, the settings are straightforward—select a voltage value, and the other parameters will adjust automatically. If the machine is not synergic, you will need to manually adjust the wire feed speed and welding current to find the right balance.
   - **Manual adjustment of the machine:**
     - o If using a non-synergic machine, start with the maximum wire feed speed and set the voltage to 15–18 volts.
     - o Open the arc on a scrap piece and gradually reduce the wire speed until the sound changes to a smooth "purring"—a sign of a well-adjusted machine.
     - o Avoid reducing the speed too much, as this can cause large droplets to form at the wire tip.

WELDING PROCEDURE

1. **Initial positioning**
   - **Welder's position:**
     - o Right-handed welders: Weld from right to left, keeping your head towards the left so that your eyes are at the end of the weld bead, ensuring clear visibility of the wire progression.
     - o Left-handed welders: Weld from left to right using the same principle.
   - **Position test:** Without pressing the trigger, practise moving the torch along the marked guidelines. Ensure you can maintain a consistent angle and distance without difficulty.

2. **Executing the weld beads**
   - **First weld bead:** Start welding by pushing the torch forward along the first marked line. Maintain a torch inclination of approximately 10° backward. Keep a consistent distance of 3/8″ to 5/8″ between the contact tip and the workpiece.

- **Constant travel speed:** Maintain a steady travel speed from the start to the end of the bead—this is essential to achieve uniform width and penetration.
- **Subsequent weld beads:** Repeat the process for the remaining eight beads, always pushing the torch forward and maintaining the same distance and speed.

EVALUATION OF THE PRACTICE

1. **Weld bead straightness**
   - The primary objective of this exercise is to produce straight weld beads, following the punched guidelines.
   - Check that all beads have uniform width, which indicates consistent travel speed.
2. **Cleaning and maintenance**
   - Ensure the welding hose remains as untwisted as possible throughout the operation to prevent wire feeding issues.
   - Clean the nozzle and internal torch components if necessary to ensure proper shielding gas protection and prevent weld contamination.

MAG WELDING WITH CARBON STEEL

PRACTICE 2. OVERLAY WELDING IN FLAT POSITION—STRAIGHT AND WEAVE BEADS PA (1G)

- **Base material:** Carbon steel plate 4″ x 4″ x 1/8″
- **Wire diameter and designation:** 0.030″ (0.8mm) ER 70S-6 (AWS A5.18–05) G 46 3 M 2Mo (EN ISO 14341-A:2008)
- **Number of weld beads:**
  o Six straight beads
  o Three weave beads, with the torch directed forward (pushing technique)
- **Welding voltage:** 15 to 18 volts
- **Wire feed speed:** 10 to 23 feet per minute
- **Stick-out length:** 3/8″ to 5/8″ (measured from the contact tip to the weld bead)
- **Gas flow rate:** 21 to 32 cubic feet per hour (CFH) of argon (85%)/CO_2 (15%) (approximately 1 CFH per 0.04″ of nozzle inner diameter)
- **Auxiliary tools:** Ruler, scriber, measuring tape, file, grinder, centre punch, and hammer (Figure 3.5)

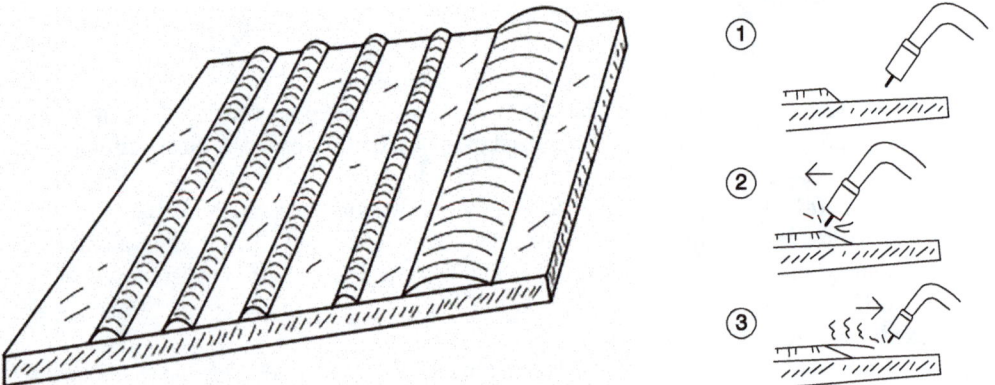

FIGURE 3.5 Metal plate with several parallel weld beads, displaying different bead sizes and profiles.

MATERIAL AND TOOL PREPARATION

1. **Measuring and cutting the plate**
   - **Measuring:** Start by measuring the carbon steel plate with a ruler or measuring tape to ensure the final dimensions are 4″ x 4″ (100 x 100 mm).
   - **Marking:** Mark the cutting line with a scriber, and make small centre punch marks along the line to guide the cut.
   - **Cutting and cleaning:**
     o   Use a grinder with a cutting disc to cut the plate along the marked lines.
     o   After cutting, remove burrs and round off the corners of the plate.
2. **Marking guide lines**
   - **Guide lines:** Mark straight lines 3/8″ apart (1cm) on the plate's surface, just as in the previous practice. These lines will serve as visual references for executing the weld beads.

EQUIPMENT SETUP

1. Adjusting Gas Flow and Wire Feed Speed
   - Set the shielding gas flow rate to 21–32 CFH (10–15L/min).
   - Adjust the wire feed speed according to the specified voltage (15–18V).
   - If using a synergic machine, select the appropriate voltage, and the machine will adjust the rest automatically.

WELDING PROCEDURE

1. **Executing straight beads**
   - **Positioning:** Align the torch with the first guide line on the plate.
   - **Restarts and overlapping welds:**
     o   During the six straight beads, stop the arc randomly and restart the weld on the same bead.
     o   When restarting, begin the new arc a few millimetres (about 1/8″) ahead of the end of the previous bead.
     o   Slowly move backward to overlap the existing weld, ensuring a smooth transition (see illustration in Figure 3.6).
     o   The goal is to avoid visible stops or discontinuities in the weld bead.
   - **Evaluating overlapping welds:**
     o   A good weld overlap should be almost imperceptible, with a smooth transition between the previous and new sections.
     o   If the overlap is visible or has a small cavity, the torch may have been misaligned or the stop too long.
2. **Executing weave beads with patterns**
   - **Weaving patterns:** Weaving consists of a side-to-side motion while advancing, allowing for wider bead coverage. Common weaving patterns include:
     o   Zig-zag: A straight-line motion with side deviations.
     o   U shape: Similar to a zig-zag, but with smoother curves.
     o   Circular: Continuous circular movements while advancing.
     o   Triangular: Triangle-shaped motion.
     o   Figure-8: Movements mimicking the number eight.

- **Pausing in simple patterns:**
  o   For zig-zag or U shape patterns changing direction.
  o   This ensures even coverage and prevents weak spots or uneven areas.
- **Continuous patterns:**
  o   For circular, triangular, or figure-8 movements, do not pause.
  o   The motion should be smooth and continuous, ensuring uniform coverage.

MAG WELDING WITH CARBON STEEL

PRACTICE 3. LEARNING TO MAKE TACK WELDS

- **Base material:** Carbon steel plate 4″ x 4″ x 1/8″
- **Wire diameter and designation:** 0.030″ (0.8mm) ER 70S-6 (AWS A5.18–05) G 46 3 M 2Mo (EN ISO 14341-A:2008)

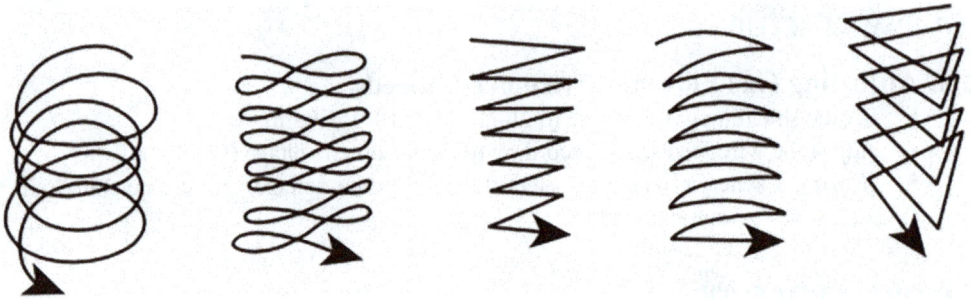

FIGURE 3.6   Illustration of six types of weaving motions used in welding, including spiral, zigzag, loop, and angular movements.

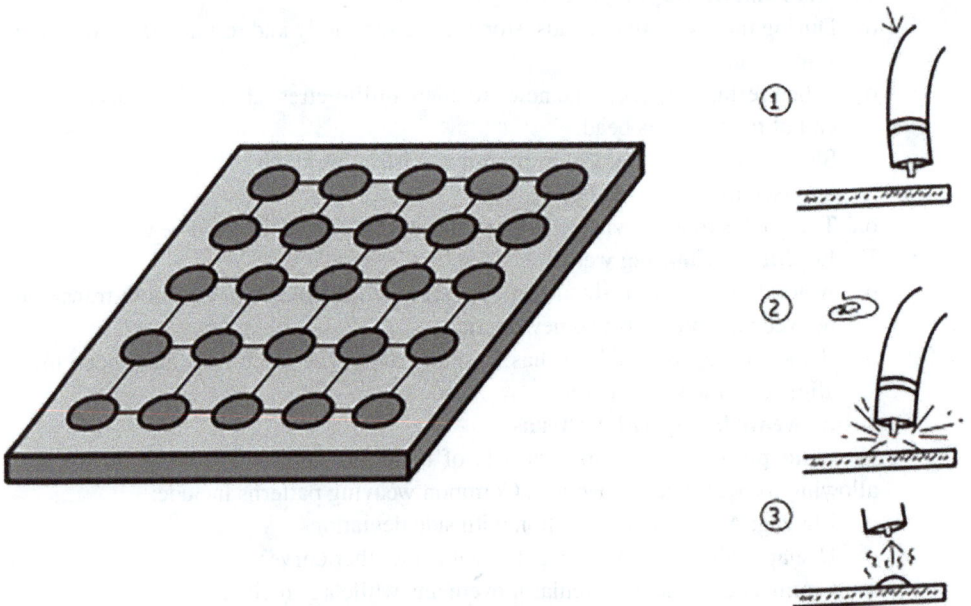

FIGURE 3.7   Metal plate with a pattern of weld spots, accompanied by three diagrams on the right showing the application sequence: electrode positioning, arc initiation, and weld spot formation.

- **Number of tack welds:** 25
- **Welding voltage:** 15 to 18 volts
- **Wire feed speed:** 10 to 23 feet per minute
- **Stick-out length:** 3/8" to 5/8" (measured from the contact tip to the weld bead)
- **Gas flow rate:** 21 to 32 cubic feet per hour (CFH) of argon (85%)/CO_2 (15%) (approximately 1 CFH per 0.04" of nozzle inner diameter, Figure 3.7)

IMPORTANCE OF TACK WELDING

Tack welding is a fundamental operation in welding. It consists of making small weld spots to hold the pieces in place during the main welding process. These tacks are essential to ensure that the pieces do not move, maintaining final structure dimensions within permitted tolerances.

Additionally, tack welds must be strong enough to support not only the weight of the structure but also the weight of the welders working on it. This makes tack welding a critical responsibility for the welder, who must perform each tack weld with precision, understanding its importance in both safety and final weld quality.

MATERIAL AND TOOL PREPARATION

1. **Measuring and cutting the plates**
   - **Measuring:** Measure and mark the carbon steel plates to ensure they have the correct dimensions of 4" x 4" (100 x 100 mm).
   - **Cutting:** Cut the plates following the marked lines, ensuring clean edges with no burrs.
2. **Surface cleaning**
   - Grinding and brushing:
     - o Clean the welding surfaces using a grinder with a flap disc to remove any rust, grease, or contaminants.
     - o Use a wire brush to further clean the surfaces, ensuring they are free of impurities before tack welding.

TACK WELDING PROCEDURE

1. **Marking the tack weld positions**
   - **Marking:** Using a scriber, draw horizontal and vertical lines on the plate, creating a grid pattern.
   - **Intersection points:** Each grid intersection represents a tack weld location. This pattern ensures uniform distribution of the tack welds across the plate.
2. **Executing the tack welds**
   - **Tack welding technique:**
     - o Position the torch over a marked intersection.
     - o Begin by tracing a small outward-to-inward spiral motion.
     - o Complete one full loop, then make a second, smaller loop inside the first.
     - o When reaching the centre of the spiral, release the trigger to stop the arc.
   - **Crater control:**
     - o It is essential to avoid craters at the end of the tack weld.
     - o Many welding machines feature a "burn back" function, which gradually stops wire feed, ensuring a smooth material deposit and preventing craters at the end of the tack weld.

EVALUATING THE TACK WELDS

1. **Weld appearance**
   - Height and shape:
     - o A good tack weld should have low height, be wide and flat, and free of craters.
     - o A small reinforcement on the back side of about 1/8″ (3mm) indicates proper penetration.
2. **Weld distribution and strength**
   - **Uniformity:**
     - o Verify that the tacks are evenly distributed according to the marked grid pattern.
     - o Ensure the distribution is adequate to support weight and structural stresses.
   - **Strength:**
     - o Check that the tack welds are strong enough to hold the pieces in place and support additional weight during the welding process.

FINAL CONSIDERATIONS

Tack welding may seem like a simple operation, but it is crucial in the welding process. These small welds not only hold pieces in place but also ensure the final structure meets quality and safety standards.

Always perform tack welds with precision—their success determines the overall quality of the welding process.

MAG WELDING WITH CARBON STEEL

PRACTICE 4. CRADLED ANGLE WELD IN HORIZONTAL POSITION PA (1F)

- **Base material:** Carbon steel plate 4″ x 1.5″ x 1/8″
- **Wire diameter and designation:** 0.030″ (0.8mm) ER 70S-3 (AWS A5.18–05) G 46 3 M 2Mo (EN ISO 14341-A:2008)

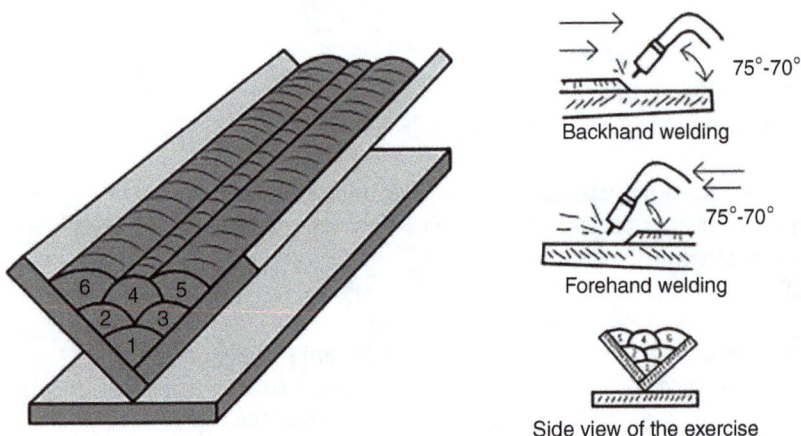

75°-70°
Backhand welding

75°-70°
Forehand welding

Side view of the exercise

FIGURE 3.8 Fillet joint with multiple weld beads, numbered to indicate the deposition sequence. On the right, diagrams show the angles and directions for backhand and forehand welding, along with a side view of the exercise.

- **Number of weld beads:** Six beads, all applied with straight motion, using either the pull (dragging) or push technique
- **Welding voltage:** 15 to 18 volts
- **Wire feed speed:** 10 to 23 feet per minute
- **Stick-out length:** 3/8″ to 5/8″ (measured from the contact tip to the weld bead)
- **Gas flow rate:** 21 to 32 cubic feet per hour (CFH) of argon (85%)/CO_2 (15%) (approximately 1 CFH per 0.04″ of nozzle inner diameter)
- **Transfer mode:** Short-circuit (Figure 3.8).

MATERIAL AND TOOL PREPARATION

1. **Measuring and cutting the plates**
   - **Measuring:** Measure the two carbon steel plates to ensure they have the correct dimensions of 4″ x 1.5″ x 1/8″ (100 x 40 x 3 mm).
   - **Marking and Cutting:** Mark the cut lines using a scriber, and use a grinder to cut the plates following the marked lines.
2. **Surface cleaning**
   - **Grinding:**
     o Use a grinder with a sanding disc to remove rust, grease, or contaminants from the surfaces to be welded.
     o Lightly round the edges to ensure a perfect fit in the angled joint.
   - **Brushing:**
     o After grinding, use a wire brush to fully clean the surfaces before welding.

ASSEMBLY AND TACK WELDING

1. **Angle assembly**
   - **Alignment:**
     o Position the two plates at a 90° angle.
     o Use a set square to confirm correct alignment.
   - **Tack welding:**
     o Tack weld the plate edges to hold them in position.
     o Ensure the tack welds are strong enough to withstand heat without shifting the plates.

WELDING PROCEDURE

1. **Executing straight beads**
   - **Weld bead sequence:**
     o This exercise involves filling the V groove with six beads, applied in a specific order:
       1. First bead: Deposited in the root of the angle.
       2. Second and third beads: Applied on either side of the first bead.
       3. Fourth bead: Applied overlapping the gap between the second and third beads.
       4. Fifth and sixth beads: Placed on either side of the fourth bead.
   - **Torch angle:**
     o Keep the torch inclined backward at about 75°–70° in the direction of travel.
     o Avoid tilting sideways, as this can compromise shielding gas coverage.

- **Travel speed:**
    - o Move the torch at a steady speed.
    - o Since you are filling a groove, move slightly slower than in previous practices to allow the wire to fill the joint properly.
- **Weaving motion:**
    - o Complement the straight motion with a slight forward-backward or side-to-side weave to ensure complete coverage of the groove.

2. **Cleaning between weld passes**
- **Removing spatter:**
    - o Between each pass, remove spatter (small metal droplets) using a hammer and wire brush.
- **Brushing:**
    - o Before starting a new weld bead, brush the surface again to ensure good fusion with the next pass.

PRACTICE EVALUATION

1. **Weld bead uniformity**
- **Alignment and filling:**
    - o Ensure each bead follows the visual guides and is evenly distributed.
    - o The weld surface should be uniform, with no irregularities or uneven areas.
- **Bead transitions:**
    - o Verify that bead overlaps are smooth, without cavities or excess material accumulation.

2. **Final cleaning**
- **Spatter removal:**
    - o Once the exercise is complete, remove all spatter from the weld surface.
- **Final inspection:**
    - o Check the entire joint to confirm there are no areas lacking material and that weld penetration is adequate.

KEY LEARNING POINTS FOR THIS PRACTICE

This practice challenges you to control travel speed and torch angle effectively while filling a cradled joint.

To achieve a uniform weld, focus on:

✓ Maintaining a steady, controlled motion
✓ Ensuring smooth overlaps between beads
✓ Applying proper torch inclination and weave motion

By mastering this exercise, you'll develop better precision in angle welding and improve your weld deposition control.

MAG WELDING WITH CARBON STEEL

PRACTICE 5. FIRST STRINGER BEADS IN HORIZONTAL VERTICAL POSITION (PC/2G)

- **Base material:** Carbon steel plate 4″ x 4″ x 1/8″
- **Wire diameter and designation:** 0.03″ (0.8mm) ER 70S-3 (AWS A5.18–05) G 46 3 M 2Mo (EN ISO 14341-A:2008)

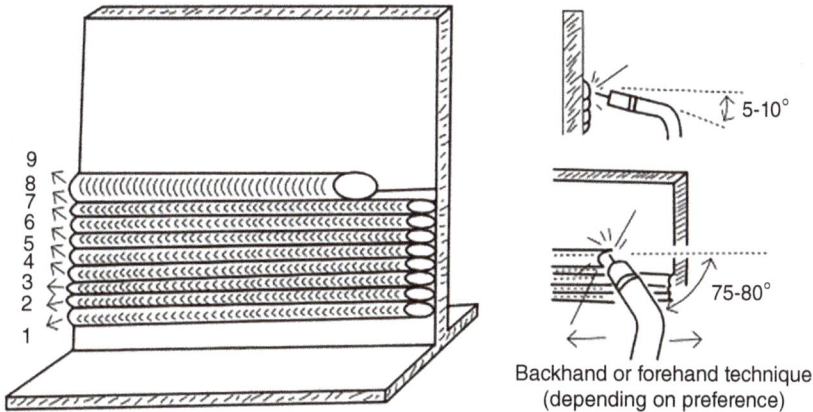

FIGURE 3.9 Welding of a plate in overhead position with multiple stacked weld beads, numbered to indicate the deposition sequence. On the right, diagrams detail the torch inclination and the option to weld either backwards or forwards.

- **Number of beads:** Approximately 20, all straight, with the gun oriented backward (dragging) or forward (pushing).
- **Welding voltage:** 15–18 volts
- **Wire feed speed:** 10–23 ft/min
- **Stick-out length:** 0.4–0.6″ (measured from the contact tip to the weld)
- **Shielding gas flow rate:** 21–32 cubic feet per hour of argon (85%)/CO_2 (15%), 1 cubic foot per inch of nozzle inner diameter
- **Transfer mode:** Short-circuit (Figure 3.9).

UNDERSTANDING THE CORNICE POSITION (PC/2G)

The cornice (PC/2G) position is a significant challenge for any welder. This is the first time you will weld in this position, allowing you to develop precise gun control, as well as work on the stability and consistency of your beads.

WELDING PROCEDURE

1. **Positioning and setup**
   - Place the steel plate in a vertical position. Ensure it is securely clamped to avoid any movement during welding.
   - Adjust the gun to the correct position:
     - o If you push the gun, tilt it forward at 75°–80° relative to the plate.
     - o If you drag the gun, tilt it backward at the same angle.
2. **Executing the straight beads**
   - **First bead:** Start welding from the lower edge of the plate, leaving a 0.4″ (1cm) gap at the bottom. Ensure the first bead is well-adhered and uniform in its length.
   - **Overlapping beads:** Move upward and ensure that each new bead overlaps the previous one by at least 50%. This overlap is crucial to guarantee good adhesion between layers and avoid defects such as lack of fusion or porosity.
   - **Controlling travel speed:**
     - o If you move too fast, the bead will be too narrow, increasing the risk of lack of fusion.

o   If you move too slowly, the bead will be too wide, increasing the risk of excessive material buildup and sagging.
o   The goal is for each bead to have the same width as the previous one and for there to be no significant variations in the weld appearance.

3. **Weld bead tie-ins**
   - It is important to stop at random points to practice weld tie-ins. When stopping, allow the bead to cool slightly before restarting the weld exactly where it ended.
   - **Restarting the bead:**
     o   When restarting, begin the new arc a few millimeters (fractions of an inch) ahead of the previous bead's endpoint.
     o   Slowly backtrack and bring the wire to the highest point of the last bead, ensuring that the heat from the new weld properly fuses the previous endpoint.
     o   This ensures a solid and uniform weld continuity.

4. **Weld inspection**
   - **Visual inspection:** After completing all 20 beads, check for consistency in bead width and height, ensuring that there is no excessive material sagging.
   - **Checking overlaps:** Each bead should be well-overlapped, with no empty spaces or lack of fusion between the layers.

RECOMMENDED GUN MOVEMENTS

1. **Straight beads**
   - Constant travel speed: Maintaining a steady travel speed is crucial.
   - The key is ensuring that each bead overlaps the previous one, maintaining a consistent width.
2. **Weave beads (optional for filling)**
   - **Small side-to-side weave:** If needed, you can use a slight side-to-side motion (up and down) to better distribute the material while advancing.
   - **Forward-backward weave:** Another useful movement is a small forward-backward weave, which concentrates heat in the desired area and ensures good fusion without excessive buildup.

FINAL CONSIDERATIONS

Welding in the PC/2G position is not only a test of your ability to control the gun and maintain a constant travel speed, but also an excellent exercise for developing the skill of overlapping beads correctly to ensure a strong and uniform weld.

Every bead you apply in this practice is an opportunity to refine your precision and consistency. Focus on maintaining a steady rhythm, ensuring proper overlap, and adjusting travel speed as necessary.

With dedication and practice, you will master this technique and be ready for more complex welding challenges in the future!

MAG WELDING WITH CARBON STEEL

PRACTICE 5. FIRST BEADS IN HORIZONTAL FILLET POSITION PC (2G)

- **Base material:** Carbon steel plate 4" x 4" x 1/8"
- **Wire diameter and designation:** 0.030" (0.8mm) ER 70S-3 (AWS A5.18–05) G 46 3 M 2Mo (EN ISO 14341-A:2008)

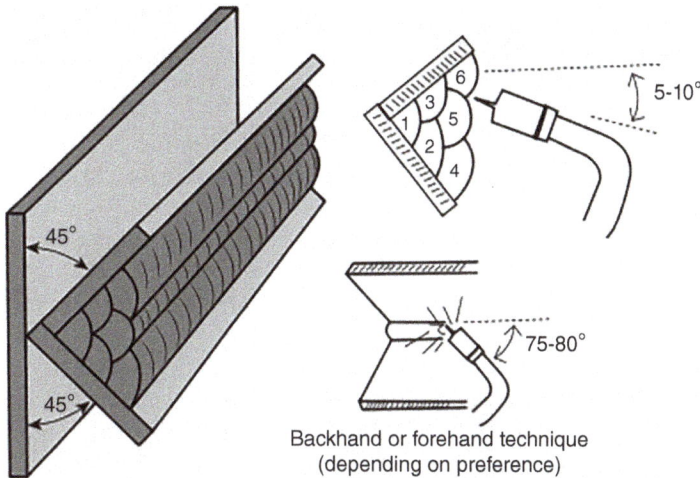

FIGURE 3.10 Illustration of a 45° fillet weld with multiple overlapping beads.

- **Number of weld beads:** Around 20, all applied with straight motion, using either the pull (dragging) or push technique
- **Welding voltage:** 15 to 18 volts
- **Wire feed speed:** 10 to 23 feet per minute
- **Stick-out length:** 3/8″ to 5/8″ (measured from the contact tip to the weld bead)
- **Gas flow rate:** 21 to 32 cubic feet per hour (CFH) of argon (85%)/CO_2 (15%) (approximately 1 CFH per 0.04″ of nozzle inner diameter)
- **Transfer mode:** Short-circuit.

WHAT IS HORIZONTAL FILLET WELDING (PC/2G)?

The horizontal fillet position (PC/2G) presents a significant challenge for any welder (Figure 3.10). This is your first experience welding in this position, which will improve your torch control, stability, and weld consistency.

WELDING PROCEDURE

1. **Positioning and setup**
   - Secure the steel plate in a vertical position. Ensure it is firmly clamped to prevent movement during welding.
   - Torch positioning:
     o If pushing, tilt the torch forward at 75° to 80° relative to the plate.
     o If dragging, tilt the torch backward at the same angle.
2. **Executing straight beads**
   - **First weld bead:**
     o Start welding at the bottom edge of the plate, leaving about 3/8″ (1cm) of space.
     o Ensure the first bead adheres well to the edge and is uniform along its entire length.
   - **Overlapping beads:**
     o Move upward for the second bead, ensuring 50% overlap with the first bead.

o   This overlap ensures proper fusion between passes and prevents defects like lack of fusion or porosity.

- **Speed control:**
    o   Keep a steady travel speed.
    o   Too fast → The bead becomes narrow with lack of fusion.
    o   Too slow → The bead becomes too wide, increasing the risk of excess material buildup or sagging.
    o   The goal is to maintain consistent bead width with no visible variations.

3. **Overlapping weld beads**
- Practice stopping and restarting weld beads at random points to perfect overlap techniques.
- **Restarting a weld bead:**
    o   Allow the previous bead to cool slightly before resuming.
    o   Start the new arc about 1/8″ (a few mm) ahead of the previous bead's end.
    o   Move backward slightly, ensuring proper heat fusion between the old and new beads.

4. **Weld inspection**
- **Visual evaluation:**
    o   After completing 20 beads, inspect them for uniform width and height.
    o   Ensure no excessive material sagging.
- **Overlap verification:**
    o   Confirm that each bead overlaps properly, with no gaps or fusion issues.

RECOMMENDED TORCH MOVEMENTS

1. **Straight beads**
- **Steady travel is essential**.
- Ensure consistent overlap between beads, maintaining the same width throughout.

2. **Weaving movements**
- **Small side-to-side weave:**
    o   If you feel the need for better material distribution, use a light lateral weaving motion.
    o   This helps prevent ridges or valleys in the weld.
- **Forward-backward weave:**
    o   A slight front-to-back motion helps concentrate heat in the fusion area.
    o   Ensures strong penetration without excessive material deposit.

FINAL CONSIDERATIONS

Welding in the horizontal fillet position (PC/2G) tests your ability to control the torch and maintain a steady movement.

Each weld bead you apply is an opportunity to refine your technique.

✓ Maintain a consistent pace
✓ Ensure proper overlap
✓ Adjust travel speed when necessary

With practice and focus, you'll master this technique and be ready for more advanced welding challenges (Figure 3.11).

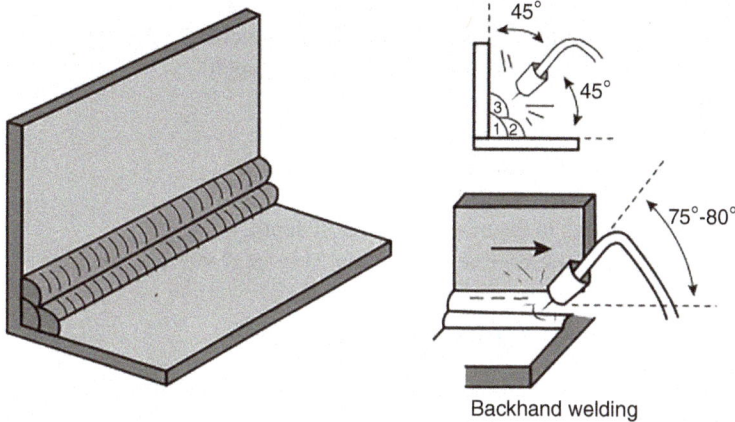

FIGURE 3.11 Fillet weld with overlapping beads in a 90° joint. On the right, diagrams show the torch angle and the backhand welding technique.

MAG WELDING WITH CARBON STEEL

Practice 7. Horizontal fillet weld pb (2f)

- **Base material:** Carbon steel plate 6" x 1.5" x 5/16"
- **Wire diameter and designation:** 0.040" (1mm) ER 70S-3 (AWS A5.18–05) G 46 3 M 2Mo (EN ISO 14341-A:2008)
- **Number of weld beads:** Three, all applied with straight motion, using either the pull (dragging) or push technique
- **Welding voltage:**
  - o 22–24 volts (short-circuit mode)
  - o 26–28 volts (spray transfer mode)
- **Wire feed speed:** 20 to 39 feet per minute
- **Stick-out length:**
  - o 3/16" to 3/8" (5–10mm) for short-circuit arc
  - o 3/8" to 5/8" (10–15mm) for spray arc
- **Gas flow rate:** 21 to 32 cubic feet per hour (CFH) of argon (85%)/CO_2 (15%)
- **Transfer mode:** Short-circuit, spray, or pulsed arc

INTRODUCTION TO THE PRACTICE

In this practice, we continue working in the horizontal position, but this time increasing the material thickness to 5/16" (8mm). This will allow us to conduct a series of experiments to observe how different welding parameters affect penetration and weld quality.

The main objective is to maximise penetration while verifying whether theoretical adjustments match practical results. At the end, we will compare outcomes to confirm whether our theories actually improve the welding process.

EXPERIMENT 1: PUSH VS. DRAG TECHNIQUE

In theory, dragging the torch (pull technique) provides deeper penetration than pushing (push technique). We will test this by welding two identical fillet joints, using the same parameters but different torch directions.

1. **Preparation**
   - Set the machine to short-circuit mode with the specified parameters.
   - Ensure the plates are clean and correctly aligned at 90°.
2. **Execution**
   - Weld the first fillet joint using the push technique.
   - Weld the second fillet joint using the drag technique.
3. **Evaluation**
   - Fracture both joints to examine penetration depth.
   - Compare results to determine if dragging truly increases penetration.

EXPERIMENT 2: ARC LENGTH VS. PENETRATION

In theory, a shorter arc length concentrates more heat at the weld root, increasing penetration. This experiment will confirm if a higher arc length reduces penetration.

1. **Preparation**
   - Set the machine to short-circuit mode with the specified parameters.
   - Adjust arc length to two levels:
     o Short arc (low arc length)
     o Long arc (high arc length)
2. **Execution**
   - Weld the first fillet joint with a short arc length.
   - Weld the second fillet joint with a long arc length.
3. **Evaluation**
   - Fracture the joints and observe how arc length affects penetration depth.

EXPERIMENT 3: STICK-OUT DISTANCE VS. PENETRATION

Stick-out distance also affects penetration. This experiment will determine whether a shorter or longer stick-out produces better results.

1. **Preparation**
   - Set the machine to short-circuit mode with the specified parameters.
2. **Execution**
   - Weld the first fillet joint with a stick-out of 3/16″ (5mm).
   - Weld the second fillet joint with a stick-out of 5/8″ (15mm).
3. **Evaluation**
   - Fracture both joints and compare penetration depth.

EXPERIMENT 4: TRANSFER MODE VS. PENETRATION (SHORT-CIRCUIT VS. SPRAY ARC)

This experiment will test whether spray transfer mode provides greater penetration than short-circuit mode.

1. **Preparation**
   - Set the machine to short-circuit mode at 22–24 volts, adjusting the wire feed speed accordingly.
   - Switch the machine to spray transfer mode at 26–28 volts, adjusting the wire feed speed again.

2. **Execution**
   - Weld the first fillet joint using short-circuit mode.
   - Weld the second fillet joint using spray transfer mode.
3. **Evaluation**
   - Fracture both joints and compare penetration depths.

FINAL EVALUATION: FRACTURING THE SPECIMENS

After welding all specimens, breaking them apart will measure penetration depth. This method is essential for welder qualification testing.

1. **Preparation for fracture**
   - Grind off the tack welds at the ends.
   - Make a deep cut along the root of the fillet joint, without cutting through the vertex.
2. **Fracturing the specimen**
   - Place the joint in a hydraulic press or bench vice.
   - Apply pressure at the ends to break the weld open.
3. **Measuring penetration**
   - Examine the fractured edge and measure the penetration depth.
   - A fully fused area will have a rough, lighter-coloured texture.
   - Use a calliper to measure penetration depth.

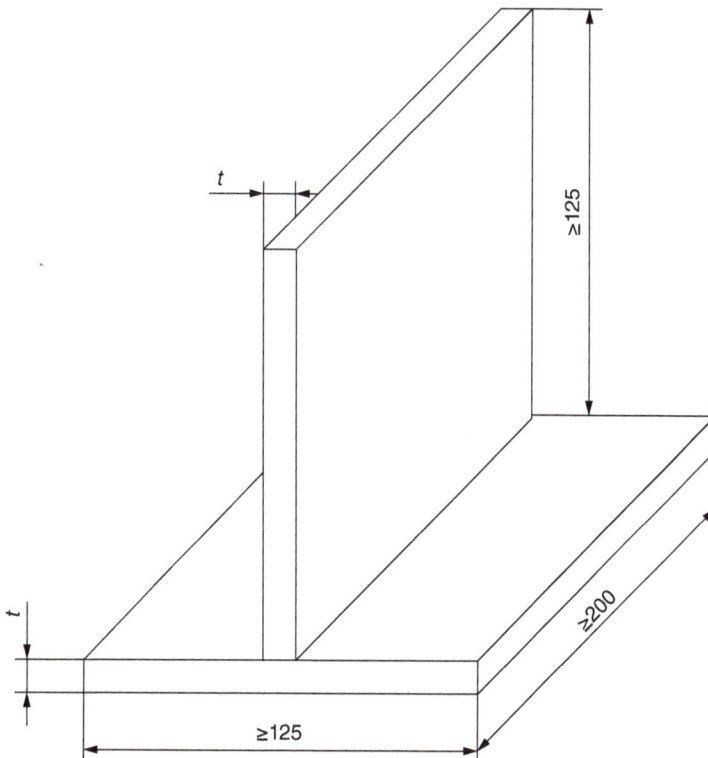

FIGURE 3.12 Technical drawing of a 90° fillet joint with specified dimensions for fabrication. The illustration depicts a vertical plate welded onto a horizontal base with minimum dimensions indicated.

CONCLUSION

If penetration is insufficient, adjust voltage and wire feed speed until achieving proper fusion. For an 5/16″ (8mm) thick plate, a penetration depth of 0.04″ (1mm) is reasonable.

The UNE EN ISO 9606–1 standard requires a minimum penetration of 0.02″ (0.5mm) across the entire weld joint.

During an official welder qualification test, you are expected to pause at least once per weld bead. Ensure that stopping points are staggered to avoid weak spots where fusion defects or porosity could develop (Figure 3.12).

If you want to practise this test on a qualification-sized coupon, the minimum dimensions required under UNE EN ISO 9606-1 are:

INSPECTION PROTOCOL FOR FILLET WELDS USING MAG IN CARBON STEEL

WELDER QUALIFICATION: PURPOSE AND IMPORTANCE

Before starting, it is essential to understand what welder qualification means and its purpose. Qualification is a process that certifies a welder's ability to perform a specific type of welding according to industry standards.

This ensures that welds meet safety, quality, and strength requirements, which is critical in industrial applications. Successfully passing a qualification test not only validates a welder's skills but also opens career opportunities in various industries.

ROLE OF THE WELDING INSPECTOR (CSWIP) IN A QUALIFICATION TEST

A Certified Welding Inspector (CSWIP) follows a strict inspection protocol when assessing welder qualification tests for fillet welds on carbon steel using the MAG process.

While the inspection steps are similar to those in SMAW (Stick) and TIG welding, certain key aspects are unique to MAG welding.

INITIAL VISUAL INSPECTION

Before welding begins
- ✓ **Joint surface condition:** The CSWIP Inspector ensures that welding surfaces are clean and free of rust, oil, or contaminants that could compromise weld quality.
- ✓ **Edge preparation:** The CSWIP Inspector verifies that the joint edges are properly prepared for a fillet weld, ensuring good fit-up and proper root clearance.

WELDING PARAMETER CONTROL

- ✓ **Equipment setup:** The CSWIP Inspector confirms that welding parameters are set according to the procedure specifications, including:
  - Voltage
  - Wire feed speed
  - Gas flow rate
  - Transfer mode (short-circuit, spray, or pulsed arc)

✓ **Shielding gas control:** The CSWIP Inspector ensures that:
  - The correct gas mixture (e.g., argon/CO_2 at the correct ratio) is being used.
  - There are no gas leaks that could affect arc stability or molten weld pool protection.

INSPECTION DURING THE WELDING PROCESS

✓ **Weld bead observation:** During welding, the CSWIP Inspector monitors the arc stability, bead consistency, and absence of surface defects, such as:
  - Porosity
  - Undercut
  - Lack of fusion
✓ **Penetration control:** While penetration is not visible during welding, the technique and parameters must be aligned with standards to ensure adequate penetration.
  - Penetration depth is later confirmed through fracture testing or X-ray inspection, if required.

FINAL INSPECTION AND DESTRUCTIVE TESTING

✓ **Final visual inspection:** After welding, the CSWIP Inspector performs a final inspection to detect:
  - Cracks
  - Surface porosity
  - Undercut or other defects
✓ **Cutting and fracture test:** The CSWIP Inspector cuts the welded joint and performs a fracture test to evaluate:
  - Penetration depth
  - Fusion between base metals
✓ **Penetration measurement:** A caliper is used to measure the root penetration, ensuring it meets industry standards (typically between 0.020″ and 0.080″ (0.5mm—2mm) depending on the applicable code).

INSPECTION REPORT AND QUALIFICATION DECISION

✓ **Final inspection report:** At the end of the test, the CSWIP Inspector compiles a detailed report including:
  - All welding parameters recorded
  - Observations on technique and execution
  - Test results (pass/fail)

This report is critical for determining whether the welder qualifies for certification.

SUMMARY

The inspection protocol for MAG fillet welds is similar to the process followed for SMAW and TIG welding.

- No major differences exist in the fundamental quality control steps.
- For additional details, refer to the sections on SMAW and TIG inspection procedures for fillet welds on carbon steel.

THE STONE AND THE SCULPTOR

In a remote Tibetan village, a sculptor was known for his breathtaking Buddha statues. Each figure radiated a peace and perfection that moved all who saw them.

A young apprentice, eager to learn the craft, approached the sculptor.

"Master, teach me how to create a statue as perfect as yours," he said.

The sculptor handed him a massive stone block and said:

"It's simple. Just chisel patiently, stroke by stroke, until you release the Buddha that is already inside the stone."

The apprentice worked for days, but the stone did not seem to take shape. Frustrated and exhausted, he returned to the sculptor:

"Master, the stone is too hard. I cannot find the Buddha within."

The sculptor smiled and replied:

"The stone is like us: it carries impurities, but within it lies perfection waiting to be revealed. Each strike not only sculpts the stone—it also sharpens your mind and strengthens your spirit. Don't give up. Keep chiseling, and both the Buddha and you will emerge transformed."

The apprentice returned to his work. With each strike, his determination grew. Over time, the statue appeared, radiant and perfect, as if it had always been waiting to be freed.

This story beautifully reflects the importance of perseverance in the learning process, even when it seems like we are not making progress. Every small effort, every mistake, and every attempt help reveal not only skill but also the greatness that resides within those who dare to try.

Want more information? Here you can watch the execution of this exercise:

Facultad de Soldadura: www.youtube.com/watch?v=8yezhSzaLgs

MAG WELDING WITH CARBON STEEL

PRACTICE 8. FIRST STRINGER BEADS + VERTICAL-UP BUILD-UP. PF(3G)

- **Base material:** Carbon steel plate 6″ x 6″ x 5/16″
- **Wire diameter and designation:** 0.04″ (1mm) ER 70S-3 (AWS A5.18–05) G 46 3 M 2Mo (EN ISO 14341-A:2008)
- **Number of beads:** Four straight beads and two weave beads, with the gun oriented forward (pushing)
- **Welding voltage:** 16–20 volts
- **Wire feed speed:** 10–23 ft/min
- **Stick-out length:** 0.2–0.4″
- **Shielding gas flow rate:** 21–32 cubic feet per hour of argon (85%)/CO_2 (15%), 1 cubic foot per inch of nozzle inner diameter
- **Transfer mode:** Short-circuit
- **Auxiliary tools:** Ruler, scribing tool, measuring tape, file, grinder, center punch, and hammer

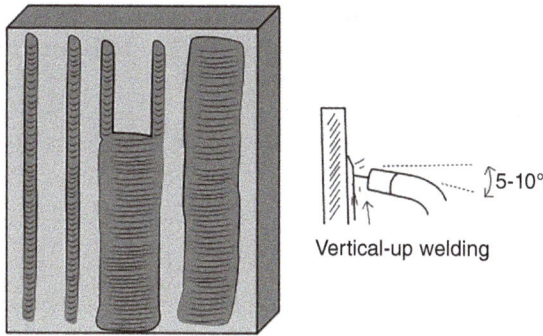

Vertical-up welding

FIGURE 3.13   Welding exercise in the vertical-up position. The image displays various weld beads applied on a plate, with differences in technique. On the right, a diagram indicates the electrode angle between 5° and 10°.

PREPARING THE WORKPIECE AND SAFETY MEASURES

In this practice, you will perform your first vertical-up welds. Working in this position requires precise control and good technique, since gravity works against you and can affect weld quality. Follow these steps carefully to complete this practice successfully (Figure 3.13).

Before starting, ensure the workpiece is clean and free of contaminants, such as rust, oil, or dirt. Use a grinder with a sanding disc to smooth the surface. Make sure the welding area is dust-free and that your equipment is in optimal condition for welding.

For safety, wear the required PPE. Specifically for this practice, it is highly recommended to wear a flame-resistant hood to protect your head and neck from spatter, along with welding gloves, a leather apron, and a welding helmet. Spatter in vertical welding is more likely to fall on you, so additional protection is essential.

WORKPIECE AND TORCH POSITIONING

- Secure the steel plate in a vertical position on the workbench, making sure it is well-clamped to prevent movement during welding.
- Maintain a torch angle of approximately 5°–10° backward from the vertical plate.

EXECUTING THE STRAIGHT BEADS

The first step is to deposit four straight vertical-up stringer beads:

- Slight zigzag motion: For these beads, make a small zigzag movement in the weld root while maintaining a constant travel speed. This will help you control the weld pool and prevent it from sagging.
- Maintain proper stick-out: Keep a stick-out length of 0.4–0.6″. If this is not done correctly, it could lead to lack of penetration or an unstable arc.

EXECUTING THE WEAVE BEADS

After completing the stringer beads, you will proceed with weave beads. Here, it's crucial to pay attention to movement patterns:

- **Zigzag or U motion:** In vertical-up welding, use a zigzag or U-shaped motion. These patterns allow you to pause briefly on the sides of the bead, ensuring even fusion and preventing porosity or lack of fusion at the edges.
- **Avoid complex patterns:** Unlike horizontal or flat welding, circular, figure-8, or triangular movements are not recommended, as they do not provide the same stability and control of molten metal.

Applying inductance theory

Inductance is a setting on your welding machine that controls how the welding current responds to sudden changes during the process. Adjusting this parameter is crucial to reducing spatter and improving arc stability.

- **High inductance:** Smoothens current response, reduces spatter, and produces a fluid, cleaner weld bead. This setting is helpful when aesthetics and smooth bead finish are important.
- **Low inductance:** Provides faster current response, which may increase spatter but improves penetration. Use this setting when penetration is more critical than bead appearance.

Inspection and quality control

Once all welds are completed, perform a visual inspection to detect any defects. A well-executed vertical-up bead should be uniform, with minimal spatter, and good penetration visible on the opposite side of the plate. If any defects are present, adjust your parameters and repeat the practice until achieving a satisfactory result.

FINAL THOUGHTS

This practice marks a significant milestone in mastering the MAG welding process: controlling the molten pool in a vertical-up position. Proper inductance adjustments will help you achieve consistent, high-quality results. Keep practicing until you feel completely comfortable in this position!

MAG WELDING WITH CARBON STEEL

Practice 9. Vertical-up fillet weld, pf (3f)

- **Base material:** Carbon steel plate 5.91″ x 1.57″ x 0.31″ (150 x 40 x 8 mm)
- **Wire diameter and designation:** 0.039″ (1 mm) ER 70S-3 (AWS A5.18–05) G 46 3 M 2Mo (EN ISO 14341-A:2008)
- **Number of beads:** Two beads (root pass and cover pass)
- **Welding voltage:** 17–21 volts (short-circuit transfer)
- **Wire feed speed:** 13–26 ft/min (4–8 m/min)
- **Stick-out length:** 0.2–0.4″ (5–10mm)
- **Shielding gas flow rate:** 21–32 cubic feet per hour (10–15 L/min) of argon (85%)/CO_2 (15%), 1 litre per millimetre of nozzle inner diameter ≈ 0.06 ft³ per inch
- **Transfer mode:** Short-circuit or spray arc

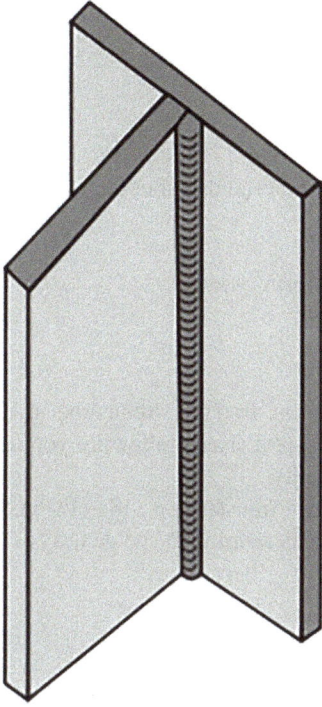

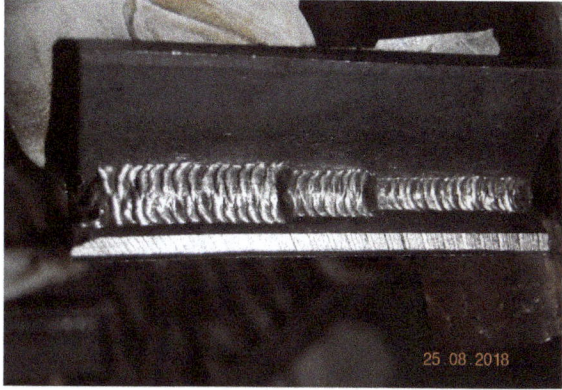

FIGURE 3.14   Three-dimensional representation of a fillet joint welded along its entire length. The image shows two plates arranged in a "T" shape with a continuous weld bead at the joint.

ROOT PASS

For this exercise, position the piece so that the highest point of the joint is at your eye level. This will allow you to have better visual control of the process and ensure that the bead remains uniform (Figure 3.14).

In this practice, we will focus on performing the root pass using different motion patterns. This pass is critical as it forms the foundation of the weld and, therefore, must penetrate properly and remain uniform.

1. **Zig-zag motion:** Make small side-to-side movements while moving upwards. This method is excellent for keeping the bead narrow while ensuring good penetration.
2. **Circular motion:** Instead of straight lines, perform small circular movements. This can make the bead slightly more rounded. If your equipment allows inductance adjustment, remember what was covered in the previous exercise: adjusting it properly can help you control the bead and reduce spatter.
3. **Triangular motion:** Form a triangle as you move upwards. Pause briefly at the lower vertices to ensure the material is evenly distributed. A useful reference: at the triangle's tip, count to one; at the lower vertices, count to three.
4. **Arrow motion:** Form an arrow, moving quickly at the tip and pausing longer on the sides. This method helps maintain a flat bead, ideal for even penetration.

COVER PASS

Once the root pass is complete, we will apply the cover pass to give the weld a final finish. Here, you can choose one of the previously mentioned patterns:

- **Zig-zag:** Ideal for ensuring full coverage on both sides of the weld.
- **U or inverted U:** These patterns provide better control over material distribution, especially when aiming for a smooth and even finish.

In both cases, make sure to pause briefly at the sides to ensure uniform coverage.

PULSED ARC

If your equipment has a pulsed arc option, this is a good opportunity to try it. Remember that pulsed arc alternates between a high current (peak) and a low current (base), allowing you to control heat input and penetration without compromising bead quality.

To set up pulsed arc, use the voltage that worked for short-circuit transfer (e.g., 18 volts) and perform a simple calculation: add and subtract 20% of that value to determine the peak and base voltages.

For our example:

- Peak voltage: 21.6 volts
- Base voltage: 14.4 volts

This setup will help maintain good heat control and minimise spatter.

SAFETY

Ensure you wear all necessary PPE to protect yourself from spatter and heat. Wear a flame-resistant hood that fully covers your head to avoid burns.

FRACTURE AND PENETRATION

The fillet weld fracture protocol, as well as the minimum penetration depth required (0.04″ or 1mm), remains the same as in the previous practice. If this penetration is not achieved, you will need to adjust the welding parameters accordingly.

CONCLUSION

This exercise will help you master vertical-up welding, one of the most challenging positions in welding. The key is to maintain steady control of the torch, correctly adjust your equipment parameters, and practice the different motion patterns until you find the one that best suits your technique and job requirements.

MAG WELDING WITH CARBON STEEL

PRACTICE 10. OVERHEAD FILLET WELD. PD (4F)

- **Base material:** Carbon steel plate 5.91″ x 1.57″ x 0.31″ (150 x 40 x 8 mm)

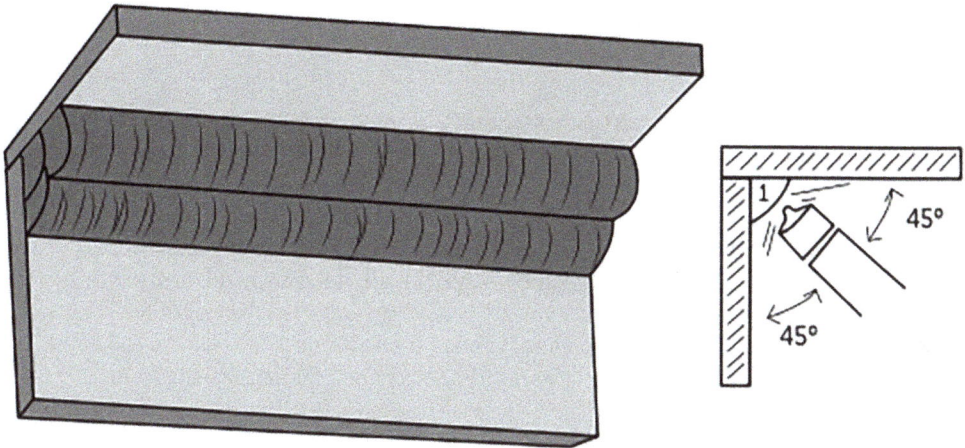

FIGURE 3.15 Illustration of a fillet weld performed in an overhead position. The main image shows two plates joined at a corner, with overlapping weld beads. On the right, a diagram indicates the electrode inclination at 45° in both directions.

- **Wire diameter and designation:** 0.039″ (1mm) ER 70S-3 (AWS A5.18–05) G 46 3 M 2Mo (EN ISO 14341-A:2008)
- **Number of beads:** Three beads, all with a straight movement, with the torch oriented either backwards (dragging) or forwards (pushing).
- **Welding voltage:** 19–23 volts (short-circuit transfer)
- **Wire feed speed:** 13–26 ft/min (4–8 m/min)
- **Stick-out length:** 0.2–0.4″ (5–10mm)
- **Shielding gas flow rate:** 21–32 cubic feet per hour (10–15 L/min) of argon (85%)/CO_2 (15%), 1 litre per millimetre of nozzle inner diameter ≈ 0.06 ft³ per inch
- **Transfer mode:** Short-circuit or pulsed arc

INTRODUCTION

This exercise, although it may seem complicated at first, is an excellent opportunity to apply everything you have learned so far, both in terms of technique and mindset. Overhead welding in the PD (4F) position requires a high level of concentration and control, but it is also an ideal scenario for learning how to manage the mental challenges of welding (Figure 3.15).

In this type of work, it is easy for the mind to anticipate failure, but it is crucial not to let these thoughts take over. If something goes wrong, take it as a learning opportunity. Analyse the mistakes, accept your limitations, and work on overcoming them with each attempt. This is the key to continuous improvement and achieving your goals.

PREPARATION AND POSITIONING

Before starting, make sure to position the piece so that the highest point of the joint is at your eye level when standing with your feet together. Then, slightly separate your feet to get a clear view of the working area while minimising the risk of burns.

Remember to use all necessary PPE, including a flame-resistant hood, to protect yourself from spatter and heat.

SHORT-CIRCUIT TRANSFER WELDING

ROOT PASS (BEAD NO. 1)

The first bead is critical. When starting, ensure that the joint line divides the weld pool into two equal parts. This is essential to achieve good penetration and prevent the bead from shifting towards the lower plate, which could cause issues in the following passes.

Move forward with a 45° torch angle relative to the joint and 80° in the direction of travel. You can use a straight movement, a back-and-forth motion, or a slight zig-zag to evenly distribute the material. Hold the torch firmly but without stiffness, allowing a smooth and steady advance.

At the end of the bead, move the wire towards the scrap tacked at the end of the plate and cut the arc. Over time, if you feel more confident, you can try finishing directly at the end of the weld, briefly pausing and then moving back a few millimetres before cutting the arc.

COVER PASSES (BEADS 2 AND 3)

The second bead should slightly overlap the first one to ensure good fusion between them. Make sure the penetration is even on both plates. For this and the third bead, you can slightly reduce the current, as the piece will be hotter.

PULSED ARC WELDING

The pulsed arc can be a great advantage in this position. It reduces the amount of material deposited, making it easier to control the weld pool and work more comfortably. Whether using a straight movement or a slight side-to-side motion, achieving a smooth and well-fused bead becomes easier.

- **Bead 1:** Set the equipment to pulsed mode. The suggested reference is as follows: take the voltage that worked well in short-circuit transfer, increase it by 20% to obtain the peak current and reduce it by 20% to get the base current. Set an arc height of -7, which will provide a more penetrating arc. Orient the torch backwards (dragging) to improve penetration and use a fine zig-zag movement, pausing only at the top of the pattern.
- **Beads 2 and 3:** Adjust the voltage following the same reference, maintaining the arc height at 0 or +7 to widen the beads. Continue with the torch oriented backwards (dragging) and the zig-zag movement, ensuring you briefly pause only at the top of the beads for a uniform finish.

FINAL CONSIDERATIONS

Remember that the fillet weld fracture test protocol and the minimum penetration depth requirement (0.02″ or 0.5mm) remain the same as in previous practices. If you do not achieve the desired penetration, adjust the parameters and try again until you reach the goal.

Work with patience and concentration, and do not be discouraged by mistakes. Every failure is an opportunity to learn and improve.

MAG WELDING WITH CARBON STEEL

PRACTICE 11. PIPE-TO-PLATE WELD IN HORIZONTAL POSITION. PB (2F)

- **Base material:** Carbon steel plate 3.94″ x 3.94″ x 0.12″ (100 x 100 x 3 mm). Pipe Ø 2.00″ x 1.38″ x 0.12″ (Ø 50.8 x 35 x 3 mm).

- **Wire diameter and designation:** 0.031–0.039″ (0.8–1mm) ER 70S-3 (AWS A5.18–05) G 46 3 M 2Mo (EN ISO 14341-A:2008).
- **Number of beads:** One with a straight movement, with the torch oriented either backwards (dragging) or forwards (pushing).
- **Welding voltage:** 18–22 volts
- **Wire feed speed:** 13–26 ft/min (4–8 m/min)
- **Stick-out length:** 0.2–0.4″ (5–10mm) for short-circuit transfer, 0.4–0.6″ (10–15mm) for pulsed arc.
- **Shielding gas flow rate:** 21–32 cubic feet per hour (10–15 L/min) of argon (85%)/CO_2 (15%), 1 litre per millimetre of nozzle inner diameter ≈ 0.06 ft³ per inch.
- **Transfer mode:** Short-circuit or pulsed arc

PRACTICE DEVELOPMENT

In this practice and the following ones, we will work with a type of joint that presents an additional challenge: one of the pieces is circular. This type of joint is common in many types of structures and requires a precise and controlled technique (Figure 3.16).

PIECE PREPARATION

Before starting, proper preparation of the pieces is essential. Make sure to remove any rust, dirt, or cutting fluid residues using a grinder with a sanding disc, a wire brush, or a finishing disc (Figure 3.17).

It is crucial that there is no gap between the pipe and the plate, as any space may compromise the quality of the weld. No light should pass through the joint at any point.

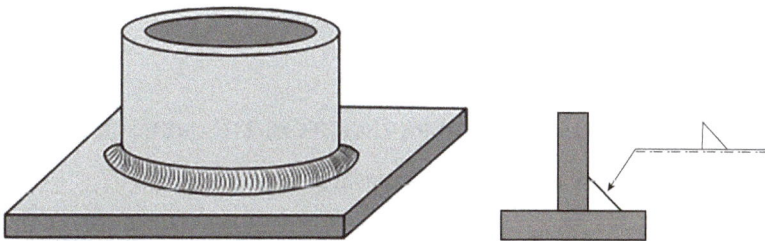

FIGURE 3.16 Illustration of a fillet weld between a cylindrical tube and a base plate. The welded joint fully surrounds the base of the tube, forming a continuous perimeter weld bead.

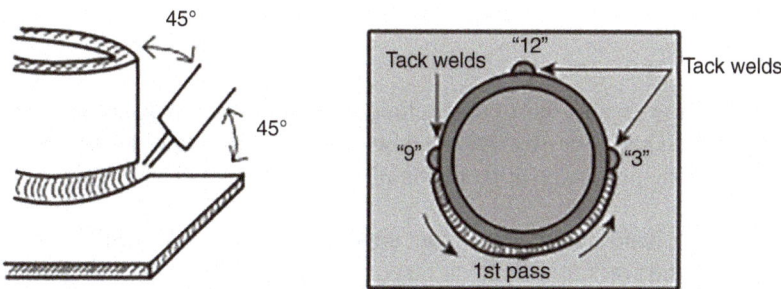

FIGURE 3.17 Diagram of a fillet weld at the joint between a cylindrical tube and a base plate. The image shows the electrode inclination angles during welding and the location of tack welds around the tube, marked with clock face references (12, 3, 6, and 9 o'clock).

TACK WELDING THE JOINT

Tack welding is essential to keep the pieces in place during welding. Apply strong and well-distributed tack welds around the pipe to prevent movement.

The tacks should be strong enough to withstand the stress and weight of the weld but should also be easy to remove with a grinder before completing the final bead.

How many tacks and where to place them? Position the pipe centrally on the plate and apply three tack welds.

POSITIONING AND WELDING TECHNIQUE

1. **Piece height:** Position the joint so that it is approximately at chest height. This height is ideal as it allows you to keep the elbow of the torch-holding arm in a comfortable and stable position without excessive elevation. This will help maintain better control of the torch movement and angle.
2. **Body positioning:** If you are right-handed, stand facing the untacked side of the joint, with the torch positioned to your right. Imagine that the pipe is the face of a clock; in this case, you will start welding at the 3 o'clock position.
3. **Torch angle:** Maintain a 75–80° inclination in the direction of travel and 45° lateral inclination. As you move forward, it is important to adjust the torch angle with a smooth wrist rotation to maintain bead consistency.
4. **Travel direction:** For this practice, start welding forwards. On a second attempt, try welding backwards, beginning from 9 o'clock to 3 o'clock.
5. **Movement and coordination:** It is essential to keep the same torch inclination throughout the weld to maintain proper gas shielding and ensure uniform penetration. Move leftward as you advance, ensuring that your head remains ahead of the torch for better visibility.

REMOVING TACK WELDS AND PREPARING FOR THE NEXT BEAD

Once welding is complete, use a grinder and a finishing disc to:

1. **Remove the tack welds:** Tack welds tend to have defects because they are applied quickly and at high current. It is important not to remelt these tacks into the main bead.
2. **Smooth the start and end of the bead:** Prepare a ramp at the start and end of the bead to facilitate starting and finishing in the next pass.

FINAL TIPS

✓ **Take your time** to get used to the technique. Run the torch along the joint without pressing the trigger, focusing only on how to adjust your wrist inclination. This will help you find the perfect position to complete both the right and left sides in a single pass.
✓ **Focus on each bead as if it were your only attempt.** This will help you concentrate and approach the practice with the necessary level of precision once you feel ready.

Want more information? Here you can watch the execution of this exercise:
Facultad de Soldadura: www.youtube.com/watch?v=wvKpeNZnpb8

MAG WELDING WITH CARBON STEEL

PRACTICE 12. PIPE-TO-PLATE WELD IN VERTICAL-UP POSITION. PH=PF FOR PIPE (2FR)

- **Base material:** Carbon steel plate 3.94″ x 3.94″ x 0.12″ (100 x 100 x 3 mm). Pipe Ø 2.00″ x 1.38″ x 0.12″ (Ø 50.8 x 35 x 3 mm).
- **Wire diameter and designation:** 0.031–0.039″ (0.8–1mm) ER 70S-3 (AWS A5.18–05) G 46 3 M 2Mo (EN ISO 14341-A:2008).
- **Number of beads:** One with a straight movement, with the torch oriented either backwards (dragging) or forwards (pushing).
- **Welding voltage:** 16–20 volts
- **Wire feed speed:** 10–26 ft/min (3–8 m/min)
- **Stick-out length:** 0.2–0.4″ (5–10mm) for short-circuit transfer, 0.4–0.6″ (10–15mm) for pulsed arc.
- **Shielding gas flow rate:** 21–32 cubic feet per hour (10–15 L/min) of argon (85%)/CO_2 (15%), 1 litre per millimetre of nozzle inner diameter ≈ 0.06 ft³ per inch.
- **Transfer mode:** Short-circuit or pulsed arc

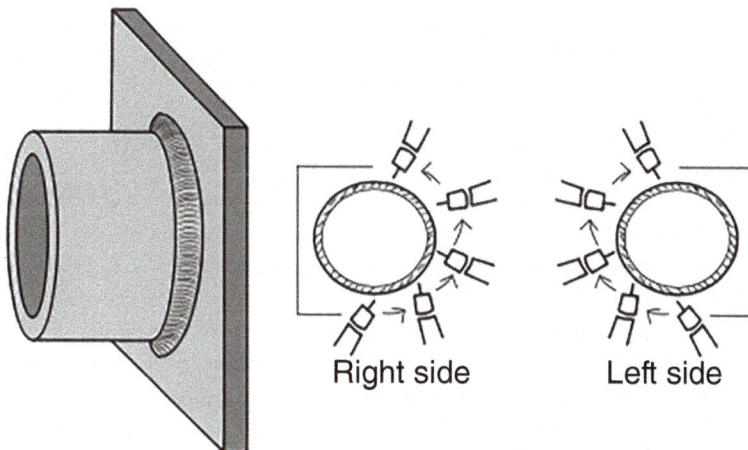

FIGURE 3.18   Fillet weld at the joint between a cylindrical tube and a base plate. The image shows the electrode inclination angles during welding and the location of tack welds around the tube, marked with clock face references (12, 3, 6, and 9 o'clock).

PRACTICE DEVELOPMENT

In this exercise, we will perform the welding of a tube to a plate in a vertical-up position (Figure 3.18). This type of joint is particularly challenging due to the need to control both the torch movement and the working angle in a confined space. Additionally, working in an upward position requires the welder to pay special attention to controlling the weld pool and ensuring proper gas shielding.

WORKPIECE PREPARATION

After properly cutting and cleaning the pieces, apply a firm tack weld. It is recommended to place three tack welds on the right side of the joint. This will ensure that the pieces remain in place during the vertical-up welding process. Make sure there is no gap between the tube and the plate, as any separation could compromise the weld quality.

POSITIONING AND WELDING

1. **Joint height:** Position the joint so that the highest point is at eye level. This will provide good visibility of the lower section of the weld seam without the need to excessively bend your torso or kneel down. Maintaining this height throughout the exercise will help control the weld bead.
2. **Body positioning:** Start welding the left side of the tube. Position yourself so that your torso is also to the left of the workpiece. Place the torch at the bottom of the tube, at the 6 o'clock position. Before starting the weld, perform a dry run by moving the torch upwards, simulating the welding movement. This will help you adjust the stick-out distance and torch angle while identifying potential difficulties before starting.
3. **Torch angle and movement:** Maintain a torch angle of approximately 75–80° in the direction of travel, adjusting the lateral inclination as needed for the bead progression. It is essential to correct the torch angle as you move, smoothly rotating your wrist to keep a consistent angle. This ensures proper gas shielding and maintains good penetration.
4. **Welding the left side:** Start welding from the 6 o'clock position. As you move up towards 10 o'clock, the torch may start obstructing your field of vision. At this point, stop, grind the end of the bead, and then reposition yourself to weld the final section with your torso on the right side of the tube. From this position, complete the weld from 10 to 12 o'clock. This approach will allow you to maintain a comfortable posture and optimal control of the torch while finishing the bead.
5. **Welding the right side:** First, remove the tack welds with a grinder, as they are no longer needed. This prevents any defects in the tack welds from being transferred to the final weld bead. Start welding from the 6 o'clock position, directing the torch forward, and move up to 2 o'clock. Then, grind the end of the bead and reposition yourself on the left side to complete the weld up to 12 o'clock.

PULSED ARC WELDING

If you decide to use the pulsed arc mode, this exercise is an excellent opportunity to see its benefits in action. Vertical-up welding with pulsed arc allows for more precise control of the weld pool and helps keep the weld cleaner. Use the same voltage and arc length adjustments described in previous exercises. For frequency adjustment, I suggest two options:

1. **Low frequency (0.5 Hz):** Move forward when the base current is active and pause when the peak current is applied.
2. **Maximum frequency:** This narrows the bead, reduces the HAZ, and increases penetration.

FINAL CONSIDERATIONS

Remember that vertical-up welding requires good technique and concentration. Maintain a comfortable posture, and do not hesitate to adjust the height or your positioning if necessary to improve control. Practice and patience are key to mastering this technique.

Additionally, if you encounter difficulties, do not get frustrated. Every attempt brings you closer to improvement and achieving a high-quality weld bead. If needed, repeat the exercise until you feel confident with the results.

MAG WELDING WITH CARBON STEEL

PRACTICE 13. OVERHEAD PIPE-TO-PLATE WELD. PD (4F)

- **Base material:** Carbon steel plate 3.94" x 3.94" x 0.12" (100 x 100 x 3 mm). pipe ø 2.00" x 1.38" x 0.12" (ø 50.8 x 35 x 3 mm).
- **Wire diameter and designation**: 0.031–0.039" (0.8–1mm) ER 70S-3 (AWS A5.18–05) G 46 3 M 2Mo (EN ISO 14341-A:2008).
- **Number of beads:** One with a straight movement, with the torch oriented either backwards (dragging) or forwards (pushing).
- **Welding voltage:** 18–22 volts
- **Wire feed speed:** 13–26 ft/min (4–8 m/min)
- **Stick-out length:** 0.2–0.4" (5–10mm) for short-circuit transfer, 0.4–0.6" (10–15mm) for pulsed arc.
- **Shielding gas flow rate:** 21–32 cubic feet per hour (10–15 L/min) of argon (85%)/CO_2 (15%), 1 litre per millimetre of nozzle inner diameter ≈ 0.06 ft³ per inch.
- **Transfer mode:** Short-circuit or pulsed arc

Although this exercise may seem intimidating at first, it is very similar to horizontal position welding. The real difficulty does not lie in the material or technique but in the psychological factor (Figure 3.19).

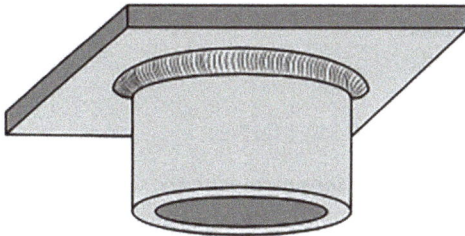

FIGURE 3.19   Overhead welded tube-to-plate joint. A peripheral weld bead is visible, securing the tube to the base.

It is common to think that the overhead position will complicate the weld finish, but it is important to understand that this mindset can interfere with your performance. The weld pool is just a molten droplet that tends to retain its shape. As long as the current is not excessive and you maintain an adequate travel speed, you should not have issues with bead sagging.

PIECE POSITIONING

The piece should be positioned at eye level.

- If placed too high, you will tire quickly and there will be a higher risk of spatter hitting your head or arms.
- If placed too low, it will be difficult to clearly see the joint.

WELDING TECHNIQUE

When welding overhead, follow the same basic recommendations:

- ✓ Find a balance to maintain a short arc length. This concentrates the arc energy in a smaller area, increasing penetration.
- ✓ Pay close attention to the stick-out length and avoid variations, as this directly affects arc length and, therefore, weld quality.

TORCH TRAVEL DIRECTION

- You can choose to weld forwards or backwards.
- In theory, welding backwards (dragging) provides greater penetration, but the best way to verify this is to experiment.

Weld a quarter of the pipe in one direction, then remove the coupon from the positioner and tear the pipe off the plate to inspect and measure penetration.

Repeat the test by welding in the opposite direction and compare the results.

Use callipers to measure the penetration depth from the inner edge to the deepest point: the weld should penetrate at least 0.04″ (1mm).

SPECIFIC CONSIDERATIONS

- ✓ Maintain a consistent torch angle from start to finish. Ensure that the wire initiates the arc at the joint and not on the bead.
- ✓ If the arc is forming on the bead instead of the joint, try the following:
  - Reduce voltage
  - Lower wire feed speed
  - Increase travel speed

MOVEMENT TESTING BEFORE WELDING

Before starting to weld, rehearse the movement with the torch without striking an arc.

- ✓ Ensure that you have a clear view of the torch's path and that you feel comfortable with your positioning.
- ✓ The torch hose can be heavy and may tend to oscillate. One option is to rest it on your shoulder, but first, check the hose condition to ensure the internal cables are well insulated.

BEWARE OF SPATTER

The nozzle tends to clog quickly in this position. It is essential to clean it frequently to prevent spatter from blocking the shielding gas flow.

Increase the gas flow rate to 32 cubic feet per hour (15 L/min) to compensate for argon's tendency to settle due to its higher density compared to air.

If, despite these precautions, porosity appears in the bead, reduce the stick-out length.

PAUSES AND ADJUSTMENTS

In the overhead position, it may be necessary to take small pauses to allow the weld pool to cool slightly, especially if you notice that it becomes difficult to control.

These pauses can also be useful for adjusting body position or torch angle if you feel that your stance or technique needs correction.

PULSED ARC WELDING

If you decide to use pulsed arc for this exercise, it will once again help you control the weld pool and reduce material deposition, which is particularly useful in this position.

As mentioned in previous exercises:

✓ Adjust the voltage to optimise the pulsed arc, using the same method of calculating 20% above and below the short-circuit voltage you used.

FINAL CONSIDERATIONS

Although it is not necessary at first, your goal should be to complete the pipe weld in two passes:

1. One pass from 12 to 6 o'clock, passing through 3 o'clock
2. One pass from 12 to 6 o'clock, passing through 9 o'clock (or vice versa)

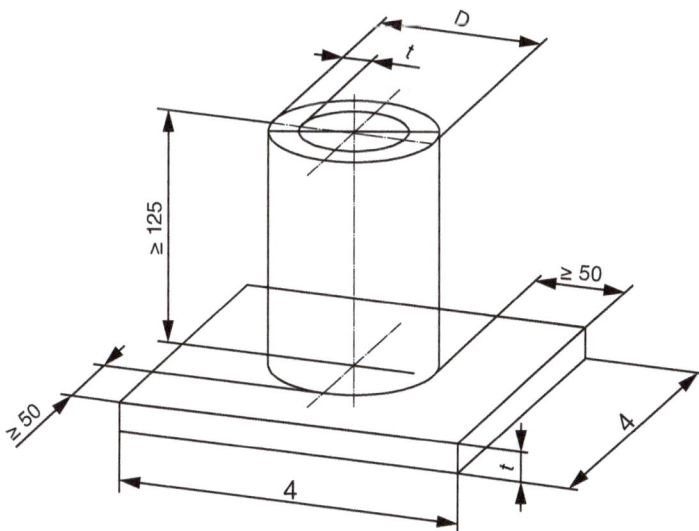

FIGURE 3.20 Welded tube-to-plate joint, with specified dimensions and measurements. The tube diameter, material thickness, and weld layout are detailed.

This exercise is excellent for developing control and refining your technique in this challenging position (Figure 3.20).

WELDING INSPECTOR (CSWIP) INSPECTION PROTOCOL FOR PIPE-TO-PLATE FILLET WELDS USING MAG ON CARBON STEEL

Regarding the inspection protocol followed by the CSWIP for certifying a welder in the MAG welding process of a test coupon consisting of a 2″ pipe with a 0.12″ (3mm) thickness welded to a 3.94″ x 3.94″ x 0.12″ (100 x 100 x 3 mm) carbon steel plate, the procedure remains virtually the same as for the TIG process.

This is because the quality criteria for weld inspection do not significantly differ between these two processes, as both must comply with the same standards for penetration, integrity, and finish. Here is a summary of the inspection protocol:

1. **Visual inspection**
   - **Detection of external defects:**
   - The CSWIP will carry out a visual inspection to identify any external defects in the weld, such as porosity, cracks, undercut, excessive convexity or concavity, misalignment, and any other surface defects.
   - **Weld finish and uniformity:**
   - The weld bead must be continuous and uniform, meeting the specified dimensions without significant deviations. Particular attention is paid to the smooth transition between the pipe and the plate.
   - **Bead measurement:**
   - The bead dimensions (width, height, and length) will be checked to ensure they fall within the permissible limits of the applicable standard. The weld must have a smooth finish and no excessive material build-up.
2. **Non-destructive testing (NDT)**
   - **Liquid penetrant or magnetic particle inspection:**
   - In some cases, liquid penetrant or magnetic particle testing may be used to detect surface defects that are not visible to the naked eye.
   - **Ultrasonic or radiographic testing:**
   - Depending on specific requirements, ultrasonic or radiographic testing may be conducted to assess penetration and identify possible internal defects such as porosity, inclusions, or lack of fusion.
3. **Destructive testing (if applicable)**
   - **Fracture test:**
   - A fracture test may be conducted on the weld to evaluate ductility and complete fusion of the joint.

In this test, force is applied to the pipe until the weld fractures, allowing the inspector to examine the internal structure of the weld and confirm root penetration and fusion.

- **Macrographic examination:**
- In some cases, a cross-section of the weld may be cut and subjected to chemical etching, revealing the internal structure of the metal.

This enables a detailed evaluation of penetration depth, grain size and distribution, and the possible presence of inclusions.

CONCLUSION

Since there are no significant differences between the inspection protocol for a MAG-welded coupon and a TIG-welded coupon under these conditions, I refer you to the chapter on the TIG process, where you can review the detailed inspection protocol at the end of the pipe-to-plate joint exercises (Figure 3.21).

MAG WELDING WITH CARBON STEEL

PRACTICE 14. BEVELLED PLATES IN V—HORIZONTAL POSITION (PA/1G)

- **Base material:** Carbon steel plate, 5.91″ x 1.57″ x 0.31″ (150 x 40 x 8 mm).
- **Wire diameter and designation:** 0.039″ (1mm), ER 70S-3 (AWS A5.18–05), G 46 3 M 2Mo (EN ISO 14341-A:2008).
- **Number of passes:** Root pass with straight movement, dragging the torch backwards. Fill and cover passes with lateral movement.
- **Welding current:**
  o Root pass: 15 to 18 volts
  o Remaining passes: 18 to 22 volts
- **Wire feed speed:**
  o Root pass: 9.8–16.4 in/min (250–420 mm/min).
  o Remaining passes: 13–29.5 in/min (330–750 mm/min).
- **Stick-out length:** 0.2–0.4″ (5–10mm) for short-circuit transfer, 0.4–0.6″ (10–15mm) for pulsed arc.
- **Gas flow rate:** 21–32ft³/h (10–15 L/min) of 85% argon/15% CO_2, approximately 2ft³/h per 0.04″ of nozzle inner diameter (1L/min per mm).
- **Transfer mode:** Short-circuit or pulsed arc.
- **Bevel angle:** 35° V bevel.
- **Additional considerations:** Root face: 0.04″ (1mm). Root gap: 0.1″ (2.5mm).

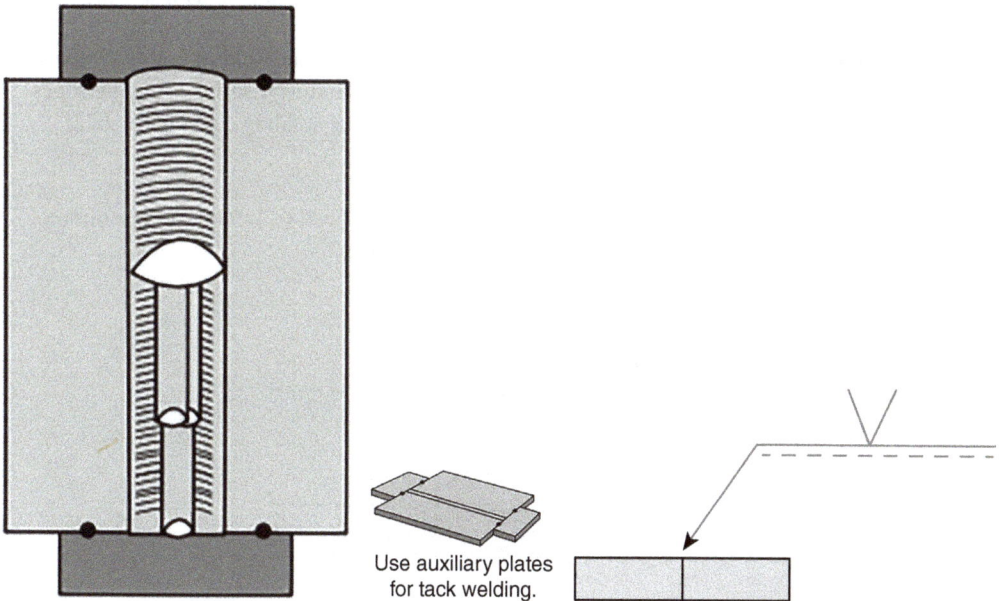

Use auxiliary plates for tack welding.

FIGURE 3.21 Butt joint weld with groove preparation, showing penetration and the sequence of weld bead deposition.

1. **Cutting and beveling:**
   o   Start by cutting the plates to the exact dimensions (150 x 40 x 8 mm) while applying a 35° V-bevel to each piece.
   o   Leave a 1mm root face on the lower edge. This prevents excessive melting of the metal, providing a better base for the root pass.
2. **Surface cleaning:**
   o   Thoroughly clean the bevelled edges and plate surfaces using a grinder and a sanding disc.
   o   Removing rust, dirt, or impurities is essential to achieving high-quality welds.
3. **Alignment and tack welding:**
   o   **Root gap:** Position the plates for welding, ensuring a root gap of 2.5mm. You can use a bare 2.5mm electrode rod bent into a V shape and placed temporarily between the plates during tacking.
   o   **Tack welding:** Tack the plates at the ends using auxiliary plates to hold them in place. Ensure strong tack welds to maintain alignment throughout the welding process.
       Root pass welding
4. **Initial setup:**
   o   Set the machine for the root pass with 15–18 volts, a wire feed speed of 3–5 metres per minute, and a stick-out length of 5–10mm.
5. **Welding technique:**
   o   **Starting:** Begin welding on the auxiliary plate and slowly drag the torch towards the root gap.
   o   **Keyhole formation:** A small hole (keyhole) should form just ahead of the weld pool, indicating proper penetration.
   o   **Adjustments:**
       1.   If no keyhole appears, stop after 2–3cm and check if the current is sufficient. Increase voltage if necessary and try again.
       2.   If the keyhole is too large, reduce its size by slightly moving the torch back and forth or using a gentle side-to-side motion. If it remains unstable, lower the voltage and wire speed.
   o   **Weld tie-in:** For a test weld, you may be required to restart the weld smoothly. Remove the last 2cm of the root bead inside the bevel using a grinding disc, ensuring the edges remain intact. This will facilitate fusion when resuming welding (see Figure 3.22).
       Filler pass welding
6. **Adjustments and settings:**
   o   Increase current to 18–22 volts and set the wire feed speed to 4–9 metres per minute.
   o   Use either short-circuit transfer or pulsed arc mode.

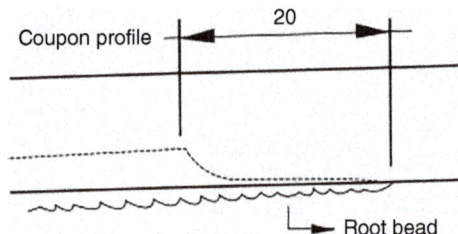

FIGURE 3.22   Diagram of a welding test specimen profile, showing the penetration of the root bead in a butt joint.

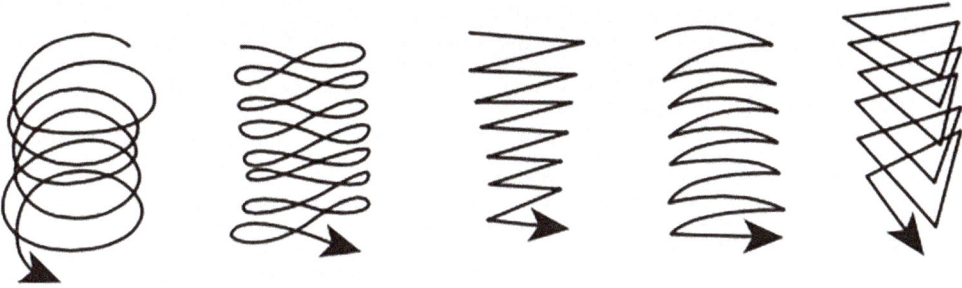

FIGURE 3.23 Illustration of six types of weaving motions used in welding, including spiral, zigzag, loop, and angular movements.

7. **Welding technique (see Figure 3.23):**
   o **Lateral movement:** Apply a gentle side-to-side motion to distribute the filler material evenly across both sides of the bevel.
   o **Objective:** Partially fill the bevel, preparing the surface for the cover pass.
     Cover (capping) pass welding
8. **Welding technique:**
   o Weaving motion: Use a slightly wider side-to-side motion following the chosen pattern to ensure complete coverage of the bevel.
9. **Final settings:**
   o Maintain the same configuration used for the filler pass, making small adjustments if necessary to improve bead appearance.
   o **Bead width:** Ensure the weld bead does not extend beyond the bevel edges. Excessive width increases the HAZ, consumes more wire, and slows down the process.
   o **Experiment 1:** Try welding with both pushing and dragging torch techniques on filler or capping passes to determine your preferred method.
   o **Experiment 2:** Adjust the arc length during the root pass to find the optimal height for controlling the keyhole.

FINAL CONSIDERATIONS

As you prepare for this MAG welding exercise, keep the following recommendations in mind to achieve high-quality results and develop essential welding skills:

1. **Body and torch positioning:**
   o **Body posture:** Maintain a comfortable and relaxed stance to ensure precise torch control. Avoid tensing your arms or body. Comfort is key to smooth, controlled movement.
   o **Torch grip:** Hold the torch firmly but without excessive stiffness. A relaxed grip allows precise movements, which is crucial for maintaining an even weld bead, especially during side-to-side motions.
2. **Visual control and focus:**
   o **Weld pool visibility:** Always keep the weld pool in sight. This is the critical area where welding occurs, and seeing it clearly helps adjust travel speed and torch angle as needed.

- o **Proper lighting:** If your work area lacks adequate lighting, adjust your position or improve illumination to ensure good visibility of the weld zone. Proper lighting is essential for detecting issues early.

3. **Pre-weld practice and simulation:**
   - o **Dry runs:** Before striking the arc, simulate torch movements to familiarise yourself with the motion, adjust your stance, and ensure good visibility and control.
   - o **Test welds:** If possible, practise on scrap metal to fine-tune your welding parameters before starting the actual weld. This is especially useful for balancing voltage, wire speed, and other settings.

4. **Heat control:**
   - o **Heat distribution:** Learn to manage heat input, especially when making multiple passes (root, filler, and capping). Overheating can cause distortion or compromise weld quality. If overheating occurs, allow the material to cool before continuing.
   - o **Strategic pauses:** If the weld pool becomes difficult to control, take short breaks to let the material cool slightly and readjust your position or technique.

5. **Learning mindset:**
   - o Accepting mistakes: Errors are part of the learning process. The key is to learn from them. Identify any issues (such as difficulty controlling the keyhole) and adjust your technique accordingly. Continuous improvement is essential for welding success.

6. **Post-weld inspection and cleaning:**
   - o Weld bead cleaning: After completing the weld, clean the bead with a wire brush to remove slag and spatter. This not only improves the final appearance but also facilitates weld inspection.

This exercise is fundamental for developing skills in full-penetration welding in the horizontal position, reinforcing techniques that will help you achieve professional-quality results.

Want more information? Here you can watch the execution of this exercise: Facultad de Soldadura: www.youtube.com/watch?v=ykLCSgshNtY

MAG WELDING WITH CARBON STEEL

PRACTICE 15. BEVELLED PLATES IN V—HORIZONTAL POSITION (PC/2G)

- **Base material:** Carbon steel plate, 5.91″ x 1.57″ x 0.31″ (150 x 40 x 8 mm).
- **Wire diameter and designation:** 0.039″ (1mm), ER 70S-3 (AWS A5.18–05), G 46 3 M 2Mo (EN ISO 14341-A:2008).
- **Number of passes:** Root pass with straight movement, dragging the torch backwards. Remaining passes with straight movement, torch either pushing or dragging.

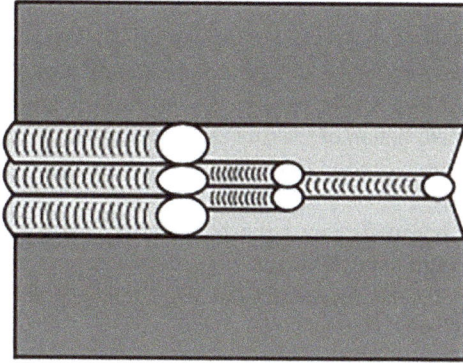

FIGURE 3.24 Butt weld on a V-groove joint, showing the deposition of multiple weld beads in successive layers.

- **Welding current:**
  - o Root pass: 15 to 18 volts
  - o Remaining passes: 18 to 22 volts
- **Wire feed speed:**
  - o Root pass: 9.8–16.4"/min (250–420 mm/min).
  - o Remaining passes: 13–29.5"/min (330–750 mm/min).
- **Stick-out length:** 0.2–0.4" (5–10mm) for short-circuit transfer, 0.4–0.6" (10–15mm) for pulsed arc.
- **Gas flow rate:** 21–32 cubic feet per hour (10–15 L/min) of 85% argon/15% CO_2, approximately 2 cubic feet per hour per 0.04" of nozzle inner diameter (1L/min per mm).
- **Transfer mode:** Short-circuit or pulsed arc.
- **Bevel angle:** 35° V bevel.
- **Additional considerations:** Root face: 0.04" (1mm). Root gap: 0.1" (2.5mm).

INITIAL PREPARATION

After performing all standard tasks of cutting, beveling, and tack welding, you can begin the actual welding process (Figure 3.24). In this position, it is crucial to apply the techniques learned in previous exercises and fine-tune the parameters according to the specific conditions of overhead welding.

BEAD 1: ROOT PASS

1. **Parameter setup:**
   - o Use the same parameters that worked well in the previous practice, adjusting the voltage between 15 and 18 volts, and the wire feed speed between 9.8–16.4"/min (250–420 mm/min).
2. **Torch angle and positioning:**
   - o The torch should be tilted between 75° and 80° backwards in the direction of travel.
   - o For lateral inclination, aim slightly more towards the bottom plate. This is key because, in overhead welding, heat tends to rise, causing the "keyhole" to form first on the upper plate.
   - o To ensure proper fusion of the lower plate, adjust the lateral inclination accordingly, and consider slightly increasing the root face of the upper plate (e.g., 0.08" (2mm) on top and 0.04" (1mm) on the bottom).

3. **Keyhole control:**
   o Controlling the size of the keyhole is essential. If it opens too wide, apply a slight oscillation motion with the torch (either forward and backward or up and down).
   o The goal is to achieve a root pass with a uniform height of approximately 0.08″ (2mm) and complete fusion of the root face. Seeing the arc light shining through the backside of the joint is a good indication of adequate penetration.
4. **Error correction:**
   o If the root pass is unsatisfactory, it must be removed using a cutting disc, taking care to maintain the original bevel intact.
   o This correction is not only necessary but also valuable practice in handling essential tools.

BEADS 2 AND 3: INITIAL FILL PASSES

1. **Parameter setup:**
   o Increase voltage to 18–21 volts and adjust wire feed speed between 13–29.5″/min (330–750 mm/min).
2. **Welding technique:**
   o Once the root pass is complete, brush the area thoroughly to remove any residue. Proceed to fill the bevel with the required number of passes.
   o Generally, a total of six passes is applied, distributed over three layers.
3. **Bead distribution:**
   o For reference, each bead should cover approximately 0.1–0.14″ (2.5–3.5mm) of the bevel width.
   o Thinner layers are preferable to avoid issues like porosity and gas inclusions, as they maintain a more consistent temperature.
   o Follow the recommended order:
      ☐ Bead No. 2 should be placed just below the root pass, pointing towards the lower line of bead No. 1.
      ☐ Bead No. 3 should be placed above, pointing towards the upper line of bead No. 1.
   o You may choose to push the torch forward in these passes.
4. **Root bead grinding:**
   o If the visible side of the root pass is not completely smooth, grind it down slightly with a finishing disc.
   o This will help ensure that beads No. 2 and No. 3 are also flat and uniform.

Beads 4, 5, and 6: Final fill and capping passes

1. **Bead 4:**
   o This pass should fill the bevel up to the marked edge.
   o It is important to maintain a steady travel speed, as the remaining space will be limited.
   o A consistent pace will ensure a flat and smooth bead without overflow.
2. **Beads 5 and 6:**
   o These passes will be laid over bead No. 4 following the same process.
   o In bead No. 5, ensure a constant width.
   o In bead No. 6, slow down the travel speed slightly to ensure full coverage of the edge line.

3. **Finalisation:**
   o   The ultimate goal is a uniform weld with as few defects as possible.
   o   Ensure each bead is properly fused and that the transitions between them are smooth.

FINAL CONSIDERATIONS

- **Trial and error:** Do not hesitate to adjust parameters if something does not go well in the first attempts. Removing a poorly executed pass and trying again is part of the learning process.
- **Practice:** Overhead welding is demanding, but with consistent practice and attention to detail, you can master it. Take the time to analyse each step and improve with each attempt.
- **Safety:** Always maintain a strong focus on safety. Regularly inspect your equipment and use the appropriate PPE.

This exercise is another step towards welding mastery. With each practice session, you will get closer to the precision and consistency expected in professional work.

THE RIGHT WAY TO LEARN

In a Tibetan village, a young man named Tenzin was eager to gain wisdom. One day, he decided to visit an old sage who lived atop a nearby mountain. After a long journey, Tenzin reached the sage's humble dwelling and expressed his desire to learn.

The old man, smiling serenely, replied:

"Before I teach you, I need you to complete a task. Take this spoon, fill it with oil, and walk through the entire village without spilling a single drop. Once you return, I will impart knowledge to you."

Tenzin accepted the challenge. Carefully, he walked through the village, focusing entirely on keeping the spoon steady and the oil intact. When he returned, he proudly showed the sage the full spoon.

The old man asked:

"As you walked, did you notice the flowers in the fields, hear the birds singing, or greet the villagers?"

Embarrassed, Tenzin replied:

"No, master. I was so focused on not spilling the oil that I paid attention to nothing else."

The sage said:

"Now, walk through the village again, but this time, observe everything around you and enjoy the journey."

Tenzin obeyed. As he walked, he admired the flowers, listened to the birds, and spoke with the villagers. Upon returning, he realised he had spilled some oil.

The sage then taught:

"Wisdom lies in balance. You must be mindful of your responsibilities without losing the ability to appreciate the world around you. Only then will you achieve true understanding."

This story reminds us that in the pursuit of knowledge, it is essential to maintain a balance between our duties and appreciating life around us.

MAG WELDING WITH CARBON STEEL

PRACTICE 16. BEVELLED PLATES IN V—VERTICAL-UP POSITION (PF/3G)

- **Base material:** Carbon steel plate, 5.91″ x 1.57″ x 0.31″ (150 x 40 x 8 mm).
- **Wire diameter and designation:** 0.039″ (1mm), ER 70S-3 (AWS A5.18–05), G 46 3 M 2Mo (EN ISO 14341-A:2008).
- **Number of passes**: Root pass with the torch dragging backwards. Remaining passes can be performed with either pushing or dragging, all using a lateral weaving motion.
- **Welding current:**
  o Root pass: 15 to 18 volts
  o Remaining passes: 18 to 22 volts
- **Wire feed speed:**
  o Root pass: 9.8–16.4″/min (250–420 mm/min).
  o Remaining passes: 13–29.5″/min (330–750 mm/min).
- **Stick-out length:** 0.2–0.4″ (5–10mm) for short-circuit transfer, 0.4–0.6″ (10–15mm) for pulsed arc.
- **Gas flow rate:** 21–32 cubic feet per hour (10–15 L/min) of 85% argon/15% CO_2, approximately 2 cubic feet per hour per 0.04″ of nozzle inner diameter (1L/min per mm).
- **Transfer mode:** Short-circuit or pulsed arc.
- **Bevel angle:** 35° V bevel.
- **Additional considerations:** Root face: 0.04″ (1mm). Root gap: 0.1″ (2.5mm).

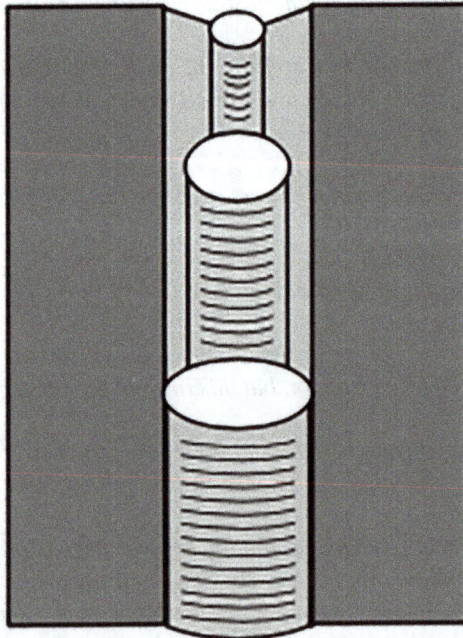

FIGURE 3.25 Butt weld on a V-groove joint, showing the deposition of multiple weld beads in successive layers.

WELDING PROCESS

PASS NO. 1: ROOT PASS

Vertical-up welding presents unique challenges, especially when executing the root pass (Figure 3.25). A common issue is that the wire may penetrate through the root gap, which can cause the arc to be interrupted, cool the weld pool, and reduce penetration quality. Additionally, protruding wire fragments may appear on the backside of the root bead.

To avoid these issues, you have two options:

1. **Welding with the torch pushing forward:**
   o   Angle the torch in the direction of travel and apply a slight lateral movement while ensuring the wire maintains continuous contact with the forming bead, preventing wire fragments from slipping through the root gap.
2. **Welding with the torch dragging backward:**
   o   This is the technique used in previous exercises and may be more suitable here, as it reduces the risk of wire slipping through the gap.
   o   In this method, the wire maintains constant contact with the forming bead as you advance.

Regardless of the technique chosen, the main objective is to control the formation of the keyhole from the beginning. Use a lateral weaving motion to stabilise it, reducing the weave if the keyhole becomes too easy to open.

PASS NO. 2: FILLER PASS

Once the root pass is complete, the next step is to cover it with the filler pass. The edges of the root pass will serve as your reference.

- Use a simple lateral movement, either a U or zigzag pattern, with a short pause at the edges to ensure even material distribution.
- The goal is to keep the bead flat, with minimal ripple marks, while leaving enough space for the final cover pass.

PASS NO. 3: COVER PASS

Finally, you must deposit the cover pass to complete the joint, particularly when working with 0.31″ (8mm) thick plates.

- Ensure you do not exceed the bevel edges to maintain a clean and professional finish.
- The final bead should fully cover the bevel without overflowing, ensuring a uniform and smooth surface.

FINAL CONSIDERATIONS

Vertical-up welding can be challenging, and it is natural to feel frustrated if the results are not perfect at first. However, every difficulty is an opportunity to learn.

- **Understand what the weld is telling you:**
  o   If the keyhole does not form properly

o If the bead deforms
o If the weld pool does not flow correctly
o Each of these is a message that, over time, you will learn to interpret and control.

THE BEST WELDING GUIDE IS YOUR OWN EXPERIENCE.
o Reflect on these questions:
   ☐ What was the most difficult part of this exercise for you?
   ☐ How did you overcome it?
o Your own insights will help you **improve and build confidence** for future welding challenges.

MAG WELDING WITH CARBON STEEL

PRACTICE 17. BEVELLED PLATES IN V—VERTICAL-UP AT 45°

- **Base material:** Carbon steel plate, 5.91″ x 1.57″ x 0.31″ (150 x 40 x 8 mm).
- **Wire diameter and designation:** 0.039″ (1mm), ER 70S-3 (AWS A5.18–05), G 46 3 M
  2Mo (EN ISO 14341-A:2008).
- **Number of passes:** Root pass with the torch dragging backwards. Remaining passes can be
  performed with either pushing or dragging, all using a lateral weaving motion (Figure 3.26).
- **Welding current:**
  o Root pass: 15 to 18 volts
  o Remaining passes: 18 to 22 volts
- **Wire feed speed:**
  o Root pass: 9.8–16.4″/min (250–420 mm/min).
  o Remaining passes: 13–29.5″/min (330–750 mm/min).
- **Stick-out length:** 0.2–0.4″ (5–10mm) for short-circuit transfer, 0.4–0.6″ (10–15mm) for
  pulsed arc.

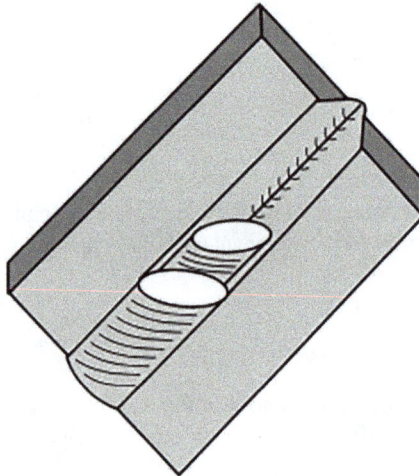

FIGURE 3.26   Joint with a 45° V-groove preparation, showing the different layers of weld beads.

- **Gas flow rate:** 21–32 cubic feet per hour (10–15 L/min) of 85% argon/15% CO_2, approximately 2 cubic feet per hour per 0.04″ of nozzle inner diameter (1 L/min per mm).
- **Transfer mode:** Short-circuit or pulsed arc.
- **Bevel angle:** 35° V bevel, with a 0.04″ (1mm) root face and a 0.1″ (2.5mm) root gap.
- **Coupon inclination:** 45° (simulating 6G position).

Pass No. 1: Root pass

The first pass, similar to the vertical-up position, presents specific challenges, but the 45° inclination introduces an additional complication: the weld pool tends to flow downward, and the wire may slip into the root gap, which could interrupt the arc and affect penetration quality.

1. **Torch position and angle control:**
   o Since the test plate is inclined, the torch angle must be adjusted to maintain a clear view of the weld pool.
   o As you move up, you will need to slightly twist the torch to the right or left (depending on the coupon's inclination) while advancing vertically.
   o Although this adjustment may feel unnatural at first, it is essential for keeping a controlled weld pool.

2. **Welding technique:**
   o It is recommended to **use the same backward welding technique** as in the previous practice, with the torch angled in the direction of travel.
   o This helps stabilise the weld pool and prevents the wire from slipping through the root gap.
3. **Lateral movement:**
   o Use a **controlled lateral weaving motion**, paying close attention to **longer pauses at the lower bevel edges** to prevent the molten material from flowing downward.
   o This weaving technique will help you **control the keyhole**, preventing it from opening excessively and losing control of the weld pool.

Passes No. 2 and No. 3: Filler and cover passes

Unlike the previous vertical-up welding practice, in this case, the filler and cover passes must be executed parallel to the ground.

This means that, when distributing the filler material, you must ensure that the torch remains in a straight line, rather than following only the coupon's inclination (Figure 3.27).

1. **Movement and material distribution:**
   o When performing filler and cover passes in this inclined position, it is important to distribute the material evenly on both sides.
   o If you weld as if it were a purely vertical position, the molten metal will tend to flow downward, causing an excess at the lower bevel edge and a lack of material at the upper edge.
2. **Material control:**
   o As you progress with the filler and cover passes, maintain a steady motion and pause slightly at the edges.
   o This will help prevent excessive material accumulation at the lower part while ensuring a uniform finish across the entire joint.

Weaving patterns for intermediate and capping passes

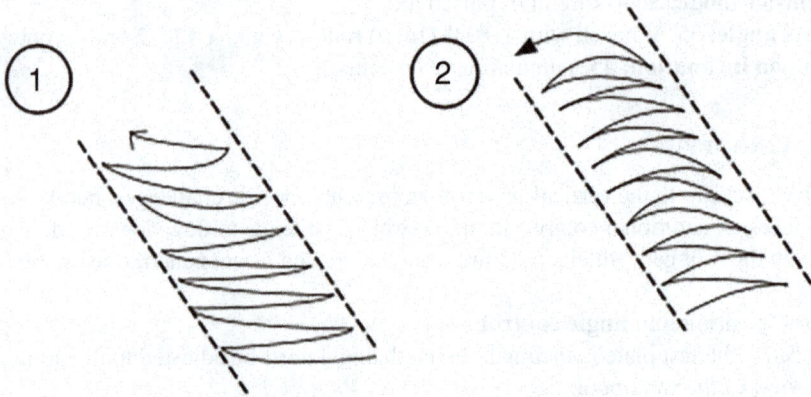

FIGURE 3.27   Illustrative diagram of welding movements for bead filling and finishing. The image displays two numbered methods (1 and 2) with lines and arrows indicating the electrode movement direction.

FINAL CONSIDERATIONS

The main difference in this practice is the adjustment of technique due to the plate's inclination.

- The 45° angle may seem confusing at first, especially since the upward movement and weaving must be combined with a slight twist to one side or the other, depending on the inclination.
- However, once you get used to these adjustments, you will be able to achieve welds as good as in standard vertical positions.
- The key is practice and familiarisation with how the molten pool behaves in this inclination, as welding in an inclined position requires greater control over heat input and weld pool stability.

MAG WELDING WITH CARBON STEEL

PRACTICE 18. BEVELLED PLATES IN V—OVERHEAD POSITION (PE/4G)

- **Base material:** Carbon steel plate, 5.91″ x 1.57″ x 0.31″ (150 x 40 x 8 mm).
- **Wire diameter and designation:** 0.039″ (1mm), ER 70S-3 (AWS A5.18–05), G 46 3 M 2Mo (EN ISO 14341-A:2008).
- **Number of passes:** Root pass with the torch dragging backwards. Remaining passes can be performed with either pushing or dragging, all using a lateral weaving motion.
- **Welding current:**
  o Root pass: 16 to 18 volts
  o Remaining passes: 18 to 22 volts
- **Wire feed speed:**
  o Root pass: 9.8–16.4″/min (250–420 mm/min).
  o Remaining passes: 13–29.5″/min (330–750 mm/min).
- **Stick-out length:** 0.2–0.4″ (5–10mm) for short-circuit transfer, 0.4–0.6″ (10–15mm) for pulsed arc.

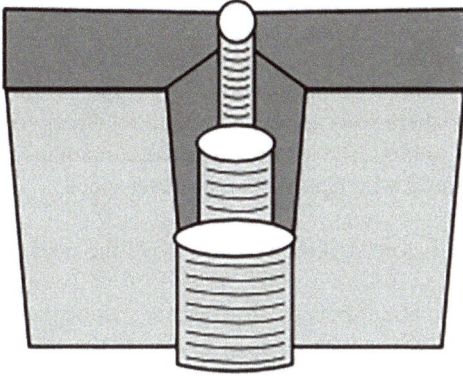

FIGURE 3.28 Overhead welding with single V preparation. The illustration depicts the joint of two bevelled pieces with multiple welding passes and a top reinforcement.

- **Gas flow rate:** 21–32 cubic feet per hour (10–15 L/min) of 85% argon/15% CO_2, approximately 2 cubic feet per hour per 0.04″ of nozzle inner diameter (1L/min per mm).
- **Transfer mode:** Short-circuit or pulsed arc.
- **Bevel angle:** 35° V bevel, with a 0.04″ (1mm) root face and a 0.1″ (2.5mm) root gap.

Overhead welding is one of the most challenging positions in welding. In this case, gravity works against you, as the molten material tends to fall before it solidifies properly (Figure 3.28).

Although this is a difficult technique, it can be mastered with a combination of practice, precision, and weld pool control (remember to wear the correct PPE).

Pass No. 1: Root pass

1. **Initial setup:**
   o Use the same voltage and wire feed speed settings that worked well in the vertical-up welding practices.
   o Make slight adjustments to match the root face and root gap of this joint.
2. **Main objective:**
   o The goal for the root pass is to break through the root face and create a keyhole without overheating the area.
   o Excessive heat can cause the penetration material to fall out, a defect that will invalidate the weld.
3. **Welding technique:**
   o Keep the torch angle at approximately 60°.
   o This will prevent the molten pool from dripping and ensure proper fusion at the root.
   o The ideal result is a flat root bead on the top side, but a slight reinforcement is even better.
4. **Root gap and root face:**
   o Start with a 0.1″ (2.5mm) root gap and a 0.04″ (1mm) root face.
   o As you gain experience, you may eliminate the root face to increase speed and improve fusion.
   o The key is to control heat input and filler material, avoiding excessive sagging.

Passes No. 2 and No. 3: Filler and cover passes

1. **Visibility and positioning:**
   o   Good visibility is critical in this position.
   o   Position yourself where you can see the weld pool clearly from start to finish.
   o   For right-handed welders, the left corner of the coupon is a good vantage point.
   o   This will help control wire placement and travel speed.
2. **Welding control:**
   o   Practice dry runs before striking the arc: move the torch along the intended path without welding.
   o   Ensure there are no blind spots or difficult areas along the travel route.
   o   If visibility is poor, adjust your position or elevate the workpiece slightly.
   o   The goal is to weld with confidence, maintaining consistent width and thickness throughout the bead.
3. **Welding technique:**
   o   Similar to horizontal welding, you can choose between:
      ☐   Two overlapping straight passes, which reduce the risk of sagging.
      ☐   A single pass with lateral movement, depending on operator skill.

Pass No. 4: Final cover pass

1. **U or zigzag motion:**
   o   For the final capping pass, use either a U or zigzag weaving motion.
   o   This movement should be steady and controlled.
   o   Start outside the joint area, on a tack-welded auxiliary plate, and gradually move forward with a slight wrist motion.
2. **Constant visibility:**
   o   Maintain clear visibility throughout the process.
   o   Good vision is essential to prevent gaps or areas lacking material.
   o   Position your head at a safe but effective distance to avoid interfering with the torch or welding helmet, while keeping a clear view of the weld pool.
3. **Clean finish:**
   o   The final weld should be smooth and free of sagging.
   o   If you followed the correct steps and controlled the travel speed, the capping pass should seal the joint cleanly and professionally.

FINAL CONSIDERATIONS

Overhead welding may seem overwhelming at first, but with patience and practice, any welder can master this technique.

- The most important factors are:
   o   Weld pool control.
   o   Maintaining a precise and steady motion.
   o   Avoiding molten metal from dripping.

Although this position is complex and often avoided when possible, mastering overhead welding will develop skills that will make you a more complete welder.

FINAL THOUGHTS

Welding constantly challenges you. The key to success is perseverance.

No one knows what they are capable of until they are faced with a difficult situation.

With practice and a positive mindset, you can overcome any obstacle in your welding journey.

Good luck and keep welding!

CSWIP WELDING INSPECTOR INSPECTION PROTOCOL FOR BUTT JOINTS IN V USING MAG ON CARBON STEEL (8MM)

The next section covers the CSWIP Welding Inspector inspection protocol specifically adapted for a welder qualification test on carbon steel coupons with 35° bevelled V joints using the MAG welding process. This protocol follows the relevant European standards and focuses on aspects specific to this type of joint.

PRE-WELDING INSPECTION

Before starting, the CSWIP Welding Inspector must verify the initial conditions of the coupon and welding equipment to ensure compliance with requirements.

- **Welding surfaces:**
  o Surfaces must be clean and free from rust, oil, and contaminants that could affect weld quality.
- **Edge preparation:**
  o The edges must be correctly bevelled to 35°, with a 0.1″ (2.5mm) root gap and a 0.04″ (1mm) root face, as specified.
- **Welding equipment settings:**
  o The CSWIP Welding Inspector will check that the welding parameters comply with the Welding Procedure Specification (WPS), ensuring appropriate voltage, wire feed speed, gas flow rate, and transfer mode.

IN-PROCESS WELDING INSPECTION

The CSWIP Welding Inspector will supervise various aspects of the welding process to ensure compliance with quality criteria.

- **Root pass:**
  o Special attention is given to keyhole control, ensuring proper penetration, which is critical in butt joints with V bevels.
  o The technique used must prevent root pass collapse and ensure proper fusion of the base material.
- **Arc stability:**
  o The CSWIP Welding Inspector will monitor arc behaviour and weld bead consistency, ensuring there are no undercuts, porosity, or unfused material inclusions.
  o In MAG welding, arc stability is crucial for producing a high-quality weld bead.
- **Penetration control:**
  o While penetration cannot be directly observed during welding, the welding parameters must be aligned with the requirements to achieve proper root fusion.
  o This will later be verified through destructive or non-destructive inspections.

POST-WELDING INSPECTION AND NDT

After completing the weld, visual inspection and, if required, additional testing will be carried out to assess internal weld quality.

- **Visual inspection:**
  - o The CSWIP Welding Inspector will conduct a visual check for external defects such as cracks, porosity, undercutting, or misalignment.
  - o The bead must be uniform in size and surface finish, without excessive convexity or concavity.
- **Liquid penetrant or magnetic particle testing:**
  - o In some cases, a penetrant or magnetic particle test may be performed to detect surface defects not visible to the naked eye.
  - o This ensures there are no cracks or surface discontinuities in the weld.
- **Radiographic or ultrasonic testing:**
  - o Depending on the qualification level and applicable standards, radiographic testing (RT) or ultrasonic testing (UT) may be performed to evaluate internal weld penetration and detect defects such as porosity, lack of fusion, or inclusions.

DESTRUCTIVE TESTING (IF REQUIRED)

Some welder qualification tests require destructive testing to verify internal weld quality.

- **Macroscopic examination (macrography):**
  - o To evaluate the internal structure of the weld, the CSWIP Welding Inspector may perform a macrography test.
  - o A section of the weld is cut and etched to reveal its internal structure.
  - o This allows assessment of:
    - ☐ Grain size
    - ☐ Penetration depth
    - ☐ Presence of inclusions

FINAL EVALUATION AND REPORT

After visual inspections and additional testing, the CSWIP Welding Inspector will compile a detailed report on the welder qualification test.

- **Inspection report:**
  - o The report must include:
    - ☐ All controlled parameters (from equipment settings to test results).
    - ☐ Observations regarding the execution of the welding process.
    - ☐ Final verdict on whether the welder has passed the qualification test.

SUMMARY

The inspection protocol for butt joints with V bevels using MAG welding follows a similar process to TIG or MMA welding but places greater emphasis on arc stability and penetration, which are critical aspects of MAG welding.

- Proper equipment parameters and root pass conditions are essential to achieve a high-quality weld.
- Final tests, both destructive and non-destructive, will ensure that the weld meets European regulatory requirements.

If the welder successfully passes all required tests, they will obtain certification for MAG welding of butt joints in V, qualifying them to perform industry-standard welds under the specified quality requirements.

WELDER QUALIFICATION TEST COUPON DIMENSIONS (ISO 9606–1)

If you wish to practise this qualification exercise using a test coupon, the minimum dimensions (in millimetres) according to UNE EN ISO 9606–1 are as follows:

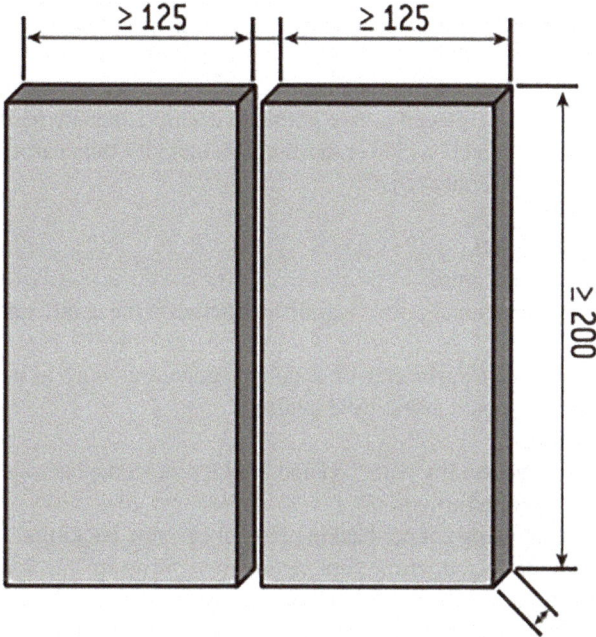

FIGURE 3.29 Technical illustration of a test coupon for butt welding on plates, displaying its minimum dimensions. Width, height, and beveling at the bottom of the material are specified.

WELDING OF AISI 304L STAINLESS STEEL WITH AWS 308L WIRE

AISI 304L stainless steel is one of the most widely used materials due to its properties, making it suitable for a wide range of applications (Figure 3.29). This type of steel is an iron-based alloy with a significant chromium content (between 12% and 30%), which gives it its oxidation resistance. Additionally, other elements such as molybdenum, titanium, and nickel enhance its tensile strength and weldability.

The welding technique for stainless steel is very similar to that used for carbon steel, but there are some additional considerations due to the specific characteristics of stainless steel and the type of electrode used.

PRECAUTIONS WHEN HANDLING AND WELDING STAINLESS STEEL

- **Cutting and preparation:**
  - o Use cutting discs and tools specifically designed for stainless steel (marked as "inox").
  - o This prevents cross-contamination with other materials, especially carbon steel.
- **Storage:**
  - o Store stainless steel separately from carbon steel and other alloys to prevent contamination.
  - o Use dedicated tools exclusively for working with stainless steel to avoid particle transfer from other materials.
- **Brushing:**
  - o Use stainless steel wire brushes for cleaning. These typically have silver bristles.
  - o Iron (golden-coloured) wire brushes are unsuitable, as they may transfer contaminants that affect the weld.
- **Tacking:**
  - o Tack welds should be placed at five points to ensure adequate fixation of the workpieces, as stainless steel has lower thermal conductivity than carbon steel.
  - o The recommended sequence is:
    1. First at the far right,
    2. Then at the far left,
    3. Followed by the centre,
    4. Finally, two additional points equidistant between the centre and the ends.
- **Cleaning:**
  - o Before welding, clean surfaces with a suitable degreaser, such as acetone, to remove contaminants that could affect weld quality.
- **Backside protection:**
  - o During welding, especially with TIG and MAG, protect the back side of the stainless steel to prevent oxide formation.
  - o This can be done using copper backing bars or ceramic backings.
  - o If not properly protected, chromium can oxidise, forming "black crystals", indicating contamination.

GAS PURGING IN STAINLESS STEEL PIPE WELDING

Gas purging is essential when welding stainless steel pipes, as it protects the inside of the weld from oxygen exposure, preventing oxidation and other defects.

- **Application of gas purging:**
  - o The purge is done by introducing inert gas (typically argon) inside the pipe before and during welding.
  - o It is essential to drill small holes at the pipe ends to allow gas flow and expel trapped air.
- **Purge time and gas flow rate:**
  - o The required purge time depends on pipe diameter and length, as well as the gas flow rate.

o A typical flow rate is 21–32 cubic feet per hour (10–15 L/min), and it must be maintained until the inside of the pipe is fully inerted.
- **Detection of oxygen presence:**
  o To ensure the pipe's interior is properly purged, oxygen detectors are used to measure concentration in parts per million (ppm).
  o An oxygen level of 20ppm or lower is generally acceptable to begin welding, as it minimises the risk of oxidation.

FORMATION OF CHROMIUM OXIDES AND CHROMIUM CARBIDES

When welding stainless steel, it is crucial to prevent the formation of chromium oxides and chromium carbides, as both can compromise corrosion resistance.

CHROMIUM OXIDES

- **Formation:**
  o Chromium oxides form when stainless steel is exposed to oxygen at high temperatures, such as during welding.
  o This appears as discolouration or a small "black cauliflower" formation on the weld surface.
- **Effect:**
  o Although a thin layer of chromium oxide is protective (as part of the passive layer in stainless steel), excessive oxidation weakens this layer, reducing corrosion resistance.
- **Prevention:**
  o To prevent chromium oxide formation, ensure:
    ☐ Proper gas purging to protect both the front and back sides of the weld.
    ☐ Minimise exposure to high temperatures.
    ☐ Use appropriate welding techniques, such as maintaining a short arc or using a faster travel speed to reduce oxidation.

CHROMIUM CARBIDES

- **Formation:**
  o Chromium carbides form when chromium reacts with carbon in stainless steel at temperatures between 842°F and 1562°F (450°C and 850°C).
  o This carbide precipitation reduces the available chromium for forming the passive layer, lowering corrosion resistance.
- **Effect:**
  o Chromium carbide formation can lead to intergranular corrosion, which attacks the grain boundaries, weakening the stainless steel structure.
- **Prevention:**
  o To prevent chromium carbide formation, follow these guidelines:
    ☐ Use low-carbon stainless steels such as 304L or 316L, specifically designed to minimise carbide precipitation.
    ☐ Control heat input during welding.
    ☐ Cool the material quickly after welding to prevent carbide precipitation.

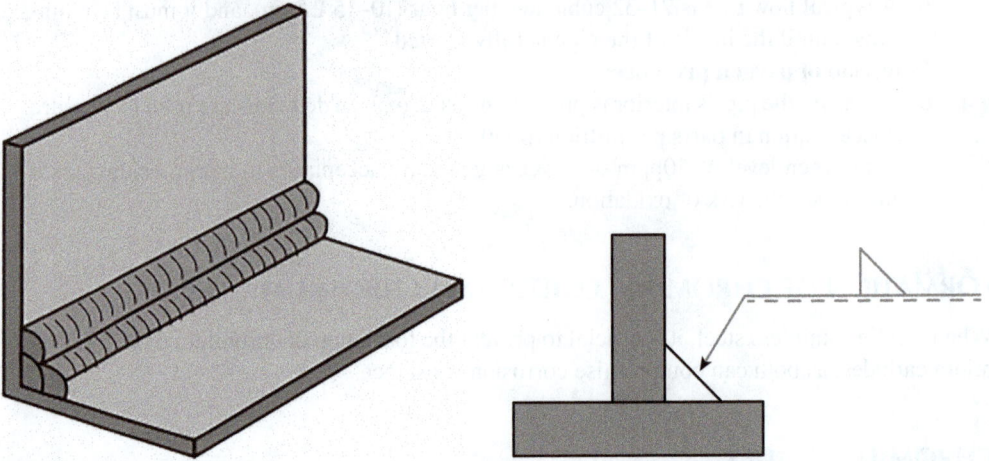

FIGURE 3.30 Illustration of a fillet weld with overlapping beads on a T-joint. Two weld beads are visible at the intersection of the perpendicular plates.

MIG WELDING WITH STAINLESS STEEL

PRACTICE 19. FILLET WELD IN HORIZONTAL POSITION—PB (2F)

- **Base material:** AISI 304l stainless steel sheet, 5.91″ x 1.57″ x 0.12″ (150 x 40 x 3 mm).
- **Wire diameter and designation:** 0.031″ (0.8mm), ER 308L (AWS A5.9), suitable for austenitic stainless steel.
- **Number of passes:** Three passes using a straight motion, with the torch pushing to improve gas shielding and avoid oxidation.
- **Welding current:** 16 to 20 volts
- **Wire feed speed:** 9.8–16.4″/min (250–420 mm/min).
- **Stick-out length:** 0.2–0.4″ (5–10mm) for short-circuit transfer, 0.4–0.6″ (10–15mm) for pulsed arc.
- **Gas flow rate:** 21–32 cubic feet per hour (10–15 L/min) of 95% argon/5% CO_2, approximately 2 cubic feet per hour per 0.04″ of nozzle inner diameter (1L/min per mm).

APPLICATION OF PULSED ARC IN THIS PRACTICE

Pulsed arc welding is an advanced MIG welding technique where the current alternates between high and low peaks (Figure 3.30). This allows molten metal to transfer in controlled droplets during the high peaks, while low current periods maintain the weld pool without overheating the material.

WHY USE PULSED ARC IN THIS PRACTICE?

When welding thin materials (less than 0.12″ or 3mm), heat control is critical to prevent burn-through or distortion. Pulsed arc welding offers the following advantages:

- **Better temperature control:** Reduces the risk of overheating stainless steel, minimising the risk of burn-through.

- **Improved penetration:** Allows consistent penetration without excessively melting the base material.
- **Reduced spatter:** Conventional MIG welding can produce excessive heat and spatter, especially with incorrect gas mixtures. Pulsed arc significantly reduces spatter, resulting in a cleaner finish.

Practical example: By using pulsed arc welding, you will notice a more stable travel speed, avoiding excessive heat accumulation and achieving more uniform weld beads on thin stainless steel sheets.

Shielding gas: Argon with a maximum of 5% CO_2

This practice uses **argon with up to 5% CO_2,** which is ideal for stainless steel welding because:

- **Minimises oxidation:**
  o Low CO_2 levels prevent excessive carbon absorption, avoiding oxidation and preserving stainless steel's corrosion resistance.
- **Enhances arc stability:**
  o Pure argon (or argon with minimal CO_2) provides a more stable and controlled arc, crucial when working with thin sheets, as it allows a smoother fusion without excessive overheating.

WHAT HAPPENS IF WE USE AN AR80/CO_2 20% MIXTURE?

If you were to use a **80% argon/20% CO_2 mixture**, commonly used for carbon steel welding, you would face the following problems:

- **Increased oxidation and reduced corrosion resistance:**
  o High CO_2 content would cause oxidation, negatively affecting stainless steel's corrosion resistance, especially in corrosive environments.
- **Poor weld finish:**
  o The weld bead would be rougher, with more spatter and thermal discolouration, impacting the visual and functional quality of stainless steel.
- **Higher risk of undercutting:**
  o A higher CO_2 percentage generates a hotter, less controlled arc, increasing the risk of burn-through and undercutting, especially on thin materials (e.g. 0.06″ or 1.5mm sheets).

To achieve a clean, stable weld that preserves corrosion resistance, always use an argon mixture with a maximum of 5% CO_2.

STEP-BY-STEP PROCESS: CUTTING, PREPARATION, TACKING, AND WELDING

1. **Cutting and preparation**
   - **Cutting the sheets:**
     o Stainless steel can be cut using a bandsaw, angle grinder with a cutting disc, or a guillotine.
     o Ensure clean and precise cuts to prevent misalignment, which can affect weld quality.

- **Cleaning the workpieces:**
  - o Remove any oxide, grease, dust, or cutting marks.
  - o Use a degreaser and stainless steel wire brush (⚠ Do not use carbon steel brushes, as they can contaminate the material!).
  - o Proper cleaning is crucial to preventing rust contamination on stainless steel.

2. **Tacking the workpieces**
   - **Alignment:**
     - o Position the sheets at a 90° angle, using a metal square to ensure accuracy.
     - o Misalignment can lead to distortion during welding.
   - **Applying tack welds:**
     - o Place three to four tacks along the joint.
     - o To avoid deformation due to stainless steel's high thermal conductivity, tack alternately:
       1. First at one end,
       2. Then at the opposite end,
       3. Finally, in the centre.
   - **Checking alignment:**
     - o Ensure the tacked workpieces remain at 90° before proceeding.

3. **Welding process**
   - **Equipment setup:**
     - o Adjust voltage and gas flow rate according to the parameters listed.
   - **Torch orientation:**
     - o Push technique (forward travel) will be used.
     - o This technique improves gas shielding and reduces oxidation, which is critical for stainless steel welding.
   - **Weld bead movement:**
     - o Use a straight-line motion for a continuous weld bead.
     - o Maintain a 45° torch angle to ensure proper fusion on both sides of the fillet weld.
   - **Arc length control:**
     - o Keep a consistent distance between the nozzle and base material.
     - o Too long an arc: Risk of lack of fusion or porosity.
     - o Too short an arc: Risk of burn-through.
   - **Travel speed:**
     - o Stainless steel heats up quickly, so maintain a steady speed to avoid overheating and distortion.

4. **Suggested parameters for pulsed arc**
   - **Voltage:** 16–20 volts (for a stable arc with good penetration and controlled heat input).
   - **Wire feed speed:** 13–20″/min (330–510 mm/min) (to ensure steady deposition without excess material).
   - **Inductance:** Medium-high (this smooths the transition between peak and base current, reducing spatter).
   - **Arc height:** 0.3–0.4″ (8–10mm) stick-out to control the weld pool and prevent instability.
   - **Pulse frequency:** Medium-high, providing a fast and stable pulsed arc, ideal for heat control on thin materials.

5. **Inspection and finishing**
   - **Visual inspection:**

  o After welding, inspect the bead. It should be uniform, with no undercuts or excessive reinforcement.

  o For stainless steel, visual quality is crucial due to its use in applications requiring high corrosion resistance and aesthetics.

- **Final cleaning:**
  - o Heat tint removal is necessary to restore stainless steel's corrosion resistance.
  - o Use either:
    - ☐ Pickling acid (for stainless steel), or
    - ☐ Electrochemical cleaning machine.
  - o This will restore the passive layer and ensure long-term corrosion resistance.

MIG WELDING WITH STAINLESS STEEL

PRACTICE 20. BUTT JOINT IN HORIZONTAL POSITION—PA (1G)

- **Base material:** AISI 304l stainless steel sheet, 5.91″ x 1.57″ x 0.06″ (150 x 40 x 1.5 mm).
- **Wire diameter and designation:** 0.031″ (0.8mm), ER 308L (AWS A5.9), suitable for austenitic stainless steel.
- **Number of passes:** One pass using a straight motion, with the torch pushing to improve gas shielding and avoid oxidation.
- **Welding current:** 14.5–17 volts for peak and base current.
- **Wire feed speed:** 9.8–16.4″/min (250–420 mm/min).
- **Stick-out length:** 0.4–0.6″ (10–15mm) for pulsed arc welding.
- **Gas flow rate:** 21–32 cubic feet per hour (10–15 L/min) of 95% argon/5% CO_2, approximately 2 cubic feet per hour per 0.04″ of nozzle inner diameter (1 L/min per mm).
- **Transfer mode:** Pulsed arc.

STEP-BY-STEP: PREPARATION, TACKING, AND WELDING A BUTT JOINT

1. **Preparation of the sheets**
   - **Precise cutting:**
     - o Cutting thin sheets (0.06″ or 1.5mm thick) requires extra care to avoid edge distortions (Figure 3.31).

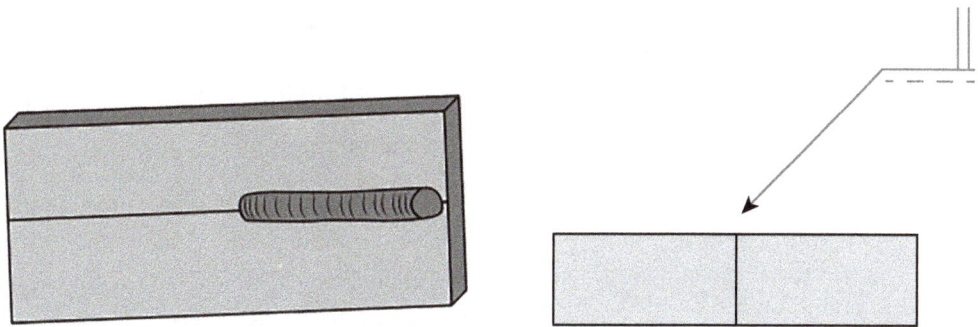

FIGURE 3.31 Technical illustration of a butt joint with a single bevel preparation, showing a weld bead applied at the root.

- o   A guillotine or an angle grinder with a thin cutting disc is recommended.
- o   Ensure that edges are properly aligned to prevent misalignment issues in the weld.
- **Material cleaning:**
  - o   Although thin sheets accumulate fewer contaminants, it is still essential to clean the surfaces properly.
  - o   Use a stainless steel wire brush or a clean cloth with a degreaser (e.g. acetone) to remove any dirt, grease, or contaminants that could affect weld quality.
- **Alignment and fit-up:**
  - o   Position the two sheets edge-to-edge, ensuring perfect alignment.
  - o   Beveling is not required for this thickness, but proper alignment is crucial to avoid deformation.
  - o   Poor alignment can result in inconsistent weld beads and distortions in the final piece.

2. **Tacking the sheets**
- **Applying tack welds:**
  - o   To prevent warping in thin 0.06″ (1.5mm) sheets, use short and spaced tack welds (approximately every 1.57″ or 40mm).
  - o   Alternate tack welds between one end and the other to distribute heat evenly and prevent distortion.
- **Proper tack weld spacing:**
  - o   In thin sheets, excessively close tack welds may cause distortions.
  - o   Evenly spaced tacks will minimise warping.

3. **Welding process**
- **Welding thin stainless steel sheets:**
  - o   Welding thin stainless steel (0.06″ or 1.5mm) requires heat control, as excessive heat can lead to deformation, burn-through, or excessive material buildup. Precise parameter adjustments are essential.
- **Current settings:**
  - o   Adjust the welding machine to a peak voltage of 17 volts and a base voltage of 14.5 volts with a high frequency pulse.
- **Welding technique:**
  - o   Hold the torch in a straight position with a minimal inclination (5–10°).
  - o   Use the push technique to improve gas coverage and reduce oxidation risk.
- **Torch movement:**
  - o   Use a smooth, controlled movement with no excessive weaving.
  - o   Avoid large oscillations, as they increase the risk of burn-through.
- **Arc and stick-out control:**
  - o   Maintain a short arc length (around 0.4″ or 10mm) to ensure proper fusion without overheating the material.
  - o   A too-long arc can cause lack of fusion or burn-through.
- **Welding sequence:**
  - o   For thin sheets, weld in short segments, allowing time for cooling between passes to avoid heat buildup.
  - o   Overheating can lead to warping, cracking, or burn-through.

4. **Cooling and finishing**
- **Controlled cooling:**

- o   Unlike carbon steel or thicker stainless steel, thin stainless sheets deform easily if cooled too quickly.
- o   Allow cooling naturally in air or use compressed air.
- o   Do not use water or liquids, as rapid cooling can cause internal stresses.
- **Final cleaning:**
  - o   Stainless steel discolours due to heat.
  - o   After welding, clean the joint with:
    - ☐   Pickling paste (for heat tint removal), or
    - ☐   Electrochemical cleaning (to restore stainless steel's original appearance and corrosion resistance).

ADDITIONAL TIPS FOR WELDING 0.06″ (1.5MM) STAINLESS STEEL SHEETS

- **Avoid overheating:**
  - o   Stainless steel heats up quickly, increasing the risk of warping.
  - o   If overheating occurs, pause and allow cooling before continuing.
- **Use clamps or fixtures:**
  - o   Thin sheets tend to warp, so use clamps or fixtures to hold them firmly in place during welding.
- **Fast but controlled travel speed:**
  - o   Advancing too slowly leads to heat buildup and warping.
  - o   Advancing too fast results in lack of fusion.
  - o   Maintain a steady, controlled speed for a smooth weld bead.
- **Use pulsed arc welding:**
  - o   Pulsed arc is highly effective for thin stainless steel, as it allows:
    - ☐   Heat control,
    - ☐   Better material transfer,
    - ☐   Reduced risk of distortion.
- **Use the correct gas mixture:**
  - o   **Argon with low CO_2 content (95% Ar/5% CO_2) ensures:
    - ☐   Minimal oxidation,
    - ☐   Stable arc,
    - ☐   High-quality welds.
- **Monitor temperature carefully:**
  - o   Stainless steel retains heat longer than carbon steel.
  - o   If distortion begins, take pauses to allow the material to cool naturally.

CONCLUSIONS

Welding **0.06″ (1.5mm) stainless steel** requires precision and heat control.

The main difference from thicker sheets is that heat cannot accumulate too much in the weld area.

Keep the arc short, maintain a steady pace, and ensure even cooling for optimal results (Figure 3.32).

If you want to practice this exercise **with a qualification test coupon**, the **minimum dimensions** according to **UNE EN ISO 9606–1** are:

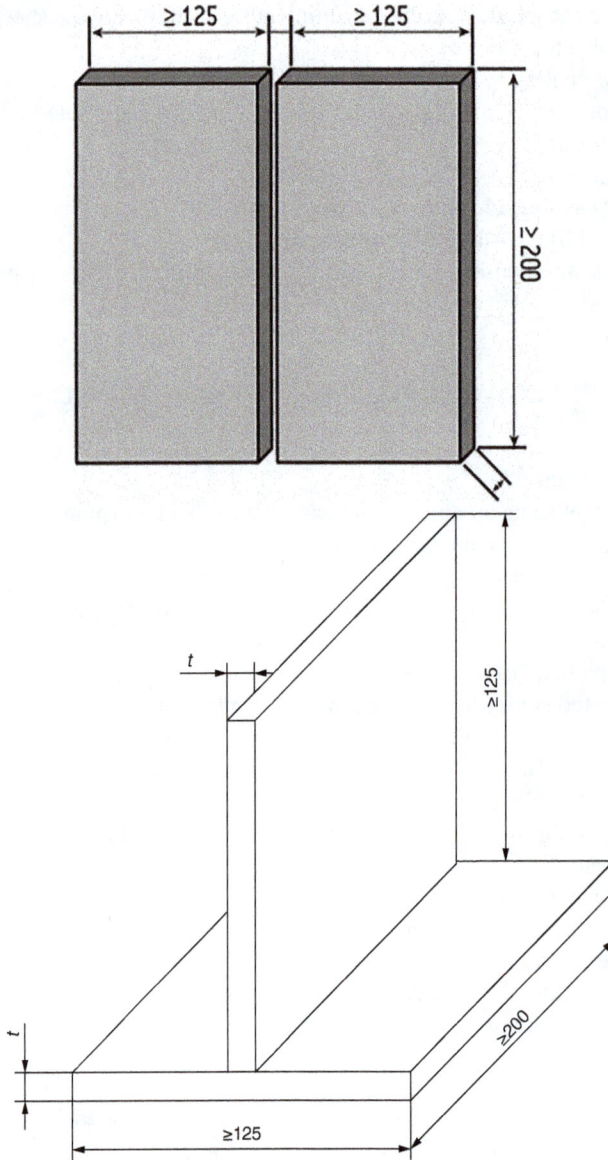

FIGURE 3.32 Technical illustrations showing the dimensions of welding test coupons. The first image depicts a butt joint between two rectangular plates with a bevel, while the second represents a fillet weld joint with a vertical plate on a horizontal base. Both images include minimum dimensions and relevant geometric details.

INSPECTION PROTOCOL FOR CSWIP WELDING INSPECTOR FOR BUTT AND FILLET WELDS USING MIG/MAG ON AISI 304L STAINLESS STEEL

To evaluate the two stainless steel AISI 304L welding test coupons (one fillet weld with 0.12″ (3mm) sheets and one butt weld with 0.06″ (1.5mm) sheets, both in a horizontal position), the CSWIP Welding Inspector must follow the ISO 9606–1 standard.

This standard defines the requirements for welder qualification and outlines the inspection and evaluation procedures for welds.

The following inspections and tests should be applied:

VISUAL INSPECTION

The first step is a visual inspection of the weld bead to verify that no surface defects are present. This applies to both the fillet and butt weld coupons.

Key aspects to verify:

- **Weld dimensions:**
  - o The weld bead must comply with the specified dimensions in the standard (width and height).
- **Weld uniformity:**
  - o The weld should be continuous, consistent, and free of interruptions.
- **Surface defects:**
  - o There must be no cracks, porosity, undercut, slag inclusions, or excessive reinforcement.
- **Fusion and penetration:**
  - o For the butt weld, ensure full fusion of the sheet edges.
  - o For the fillet weld, verify proper root fusion.

LIQUID PENETRANT TESTING (LPT)

For stainless steel welds, **non-destructive testing (NDT)** such as **liquid penetrant testing (LPT)** is recommended to detect **cracks, surface porosity, or hidden inclusions**.

This test applies to both coupons.

Procedure:

1. Apply a liquid penetrant over the weld surface.
2. Allow penetration time, then clean off the excess and apply a developer to highlight any potential defects.
3. The CSWIP Welding Inspector will check for discontinuities on the weld surface.

DIMENSIONAL MEASUREMENT

According to ISO 9606–1, precise weld bead measurements must be performed:

- **Fillet welds on 0.12″ (3mm) sheets:**
  - o The inspector will verify length, height, and width of the bead.
  - o Check for undercut or excessive reinforcement.
  - o Ensure that the 90° angle is maintained throughout the welding process.
- **Butt welds on 0.06″ (1.5mm) sheets:**
  - o The inspector will check weld thickness and confirm no excessive penetration or lack of fusion.
  - o Full root penetration and uniform bead dimensions must be achieved.

DESTRUCTIVE TESTING: BEND TEST (IF APPLICABLE)

To evaluate ductility and internal weld quality, a **bend test** may be required.

This is particularly relevant for butt welds, where penetration and fusion are critical.

Procedure:

1. Extract a sample from the welded coupon (test piece).
2. Perform a root bend test or side bend test to check for ductility.
3. The CSWIP Welding Inspector will verify no internal defects such as cracks or slag inclusions.

MACROSCOPIC EXAMINATION (MACRO TEST)

This test involves cutting a cross-section of the weld to examine the internal structure of the joint.

Macro testing helps assess penetration, filler material distribution, and internal defects.

- **Fillet welds on 0.12″ (3mm) sheets:**
  o The test will confirm bead continuity and proper root fusion.
- **Butt welds on 0.06″ (1.5mm) sheets:**
  o The test will assess full root penetration and uniform bead structure.

FINAL INSPECTION AND EVALUATION ACCORDING TO ISO 9606–1

The CSWIP Welding Inspector will complete the process by evaluating all test results.

Any defects will be compared against the acceptance criteria specified in ISO 9606–1.

ACCEPTANCE CRITERIA FOR THE WELD COUPON

- For a weld to pass, it must comply with ISO 5817 Level B or C quality criteria (depending on client requirements).
- Defects such as cracks, porosity, slag inclusions, or lack of penetration will result in rejection under ISO 9606–1.

CONCLUSION

The **CSWIP Welding Inspector's** protocol for **stainless steel welding coupons** includes:

- ☑ **A thorough visual inspection,**
- ☑ **Non-destructive testing (LPT),**
- ☑ **Dimensional checks,**
- ☑ **Destructive testing (if required),**
- ☑ **Macroscopic examination.**

Final acceptance **depends on compliance** with **ISO 9606–1** and **ISO 5817** welding quality levels.

"Strength does not come from physical capacity, but from an indomitable will."
—Mahatma Gandhi

READ THIS BEFORE CONTINUING

In the following exercises, we will focus on MIG welding of aluminium, using the 5086 series, an aluminium-magnesium alloy known for its lightweight and high strength, commonly used in the marine, aerospace, and automotive industries.

ALUMINIUM AND ITS SPECIFIC CHALLENGES IN MIG WELDING

Aluminium presents several challenges compared to other materials. Similar to TIG welding, **aluminium oxide (alumina)** is a recurring issue. This protective layer has a much higher melting point than aluminium, so its removal is critical to avoid welding defects.

CURRENT AND PULSED ARC IN MIG WELDING OF ALUMINIUM

Using pulsed arc transfer in the MIG process offers several advantages, especially for aluminium alloys. This technique allows for better molten metal transfer, reducing the risk of porosity and providing greater heat control.

This heat control is essential to prevent thermal distortion in aluminium, which heats up quickly and may warp. Additionally, pulsed arc is ideal for out-of-position welding, preventing molten metal from sagging and improving weld quality from all angles.

THERMAL CONDUCTIVITY OF ALUMINIUM AND HEAT CONTROL

Aluminium has very high thermal conductivity, meaning it dissipates heat rapidly. Unlike other materials, you will need higher amperage to maintain a proper weld pool. However, overheating is still a concern.

This is where pulsed arc transfer provides a significant advantage, allowing for controlled heat cycles.

IMPORTANCE OF CLEANING AND SHIELDING GASES

Material preparation is key. Clean the parts with a degreaser or acetone, followed by a stainless steel brush to remove the aluminium oxide layer.

It is crucial to use pure argon or an argon-helium mix (if deeper penetration is required). CO_2 mixtures cannot be used, as they cause oxidation and porosity, affecting weld quality.

COMPARISON OF SHIELDING GAS MIXTURES

- √ **100% argon:** The standard choice for aluminium, providing good penetration and oxidation protection.
- √ **Argon/helium mix:** Enhances penetration in thicker welds, but requires adjustments to welding parameters due to helium's higher thermal conductivity.

SAFETY IN ALUMINIUM WELDING

When working with aluminium, ensure that your work environment is free of aluminium dust, which is flammable and toxic.

- √ Wear protective clothing and an appropriate respirator.
- √ Ensure good ventilation when welding.

FILLER MATERIAL: 5356 VS 4043—WHAT TO EXPECT?

Choosing the right filler material is essential for achieving a strong and defect-free weld. The two most common filler wires for aluminium welding are 5356 and 4043. Here's a simple breakdown to help you choose the right one, avoiding issues like porosity and cracking.

5356 ALLOY

What is it?

5356 is a magnesium-containing filler wire. It is known for its high strength and excellent corrosion resistance, especially in marine or humid environments.

When to use it?

If welding aluminium alloys that also contain magnesium (such as 5xxx series), 5356 is a good choice.

It is ideal if the welded part will be anodised (a process that protects the surface and enhances colour), as 5356 retains colour better after anodisation.

Advantages and disadvantages

✓ **Advantage:** Lower porosity in the weld.
✗ **Disadvantage:** Slightly harder to work with compared to 4043 in some cases.

4043 ALLOY

WHAT IS IT?

4043 is a silicon-containing filler wire. It has a lower melting point, meaning it melts more easily and provides better flow during welding.

WHEN TO USE IT?

Suitable for welding aluminium alloys containing silicon (such 6xxx series).

Best for applications where good flow and gap filling are needed.

Not ideal for anodising, as it may cause colour inconsistencies in the final finish.

ADVANTAGES AND DISADVANTAGES

✓ **Advantage:** Lower risk of cracking, especially when welding crack-sensitive alloys.
✗ **Disadvantage:** More prone to porosity if the welding process is not carefully controlled.

HOW TO CHOOSE BETWEEN 5356 AND 4043

✓ **To avoid porosity:**
 • If corrosion resistance and a good anodised finish are important, use 5356 but ensure thorough surface cleaning and proper shielding gas.
 • If anodising is not required and flowability is the priority, 4043 is the better option, just ensure cleanliness to avoid porosity.

✓ **To prevent cracking:**
- 4043 is the better choice for avoiding cracks, particularly in alloys that are prone to cracking during welding.
- 5356 is more likely to cause cracks as it is less fluid and ductile than 4043.

CONCLUSION

✓ **For strength and durability in harsh environments, 5356** is the best option.
✓ For smoother welding and lower crack risk, 4043 is the better choice.

Your choice **depends on the aluminium alloy** you are welding and the **final result** you wish to achieve (Figure 3.33).

MIG WELDING WITH 5086 ALUMINIUM

PRACTICE 21. FILLET WELD IN HORIZONTAL POSITION. PB (2F)

- **Base material:** 5086 aluminium sheet, dimensions: 5.91″ x 1.57″ x 0.12″ (150 x 40 x 3 mm).
- **Wire diameter and designation:** 0.047″ (1.2mm) wire, ER 5356 or 4043 (AWS A5.10).
- **Number of beads:** One bead with a straight movement, with the torch oriented forwards (pushing) to achieve better gas shielding and prevent oxidation in the weld.
- **Welding voltage:** 16–18 volts
- **Wire feed speed root pass:** 10–16 ft/min (3–5 m/min).
- **Stick-out length:** 0.2–0.4″ (5–10mm) for short-circuit arc, 0.4–0.6″ (10–15mm) for pulsed arc.
- **Shielding gas flow rate:** 21–32 cubic feet per hour (10–15 L/min) of pure argon, 1 litre per millimetre of nozzle inner diameter ≈ 0.06 ft³ per inch.
- **Transfer mode:** Short-circuit or pulsed arc.

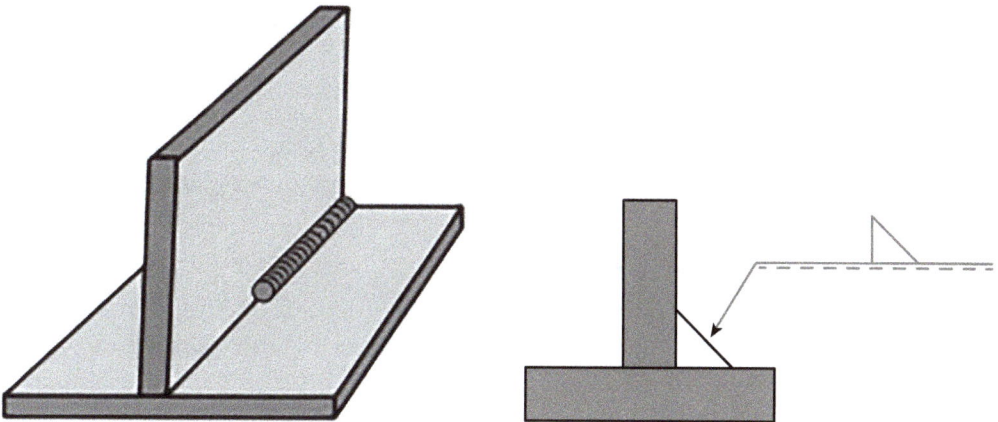

FIGURE 3.33   Technical illustration of a fillet weld.

MATERIAL PREPARATION

5086 aluminium, containing magnesium, is known for its good corrosion resistance and high weldability. Proper preparation is crucial to prevent contamination and weld defects.

- ✓ **Cutting the pieces:** You can cut aluminium using a bandsaw or a guillotine shear. Ensure that cuts are precise and uniform to achieve proper alignment during welding.
- ✓ **Cleaning the pieces:** Use a suitable degreaser or acetone to remove any traces of grease or oil. Additionally, scrub the surface with a stainless steel brush to remove the aluminium oxide layer (alumina) just before welding. Ensure that no insulating layer remains on the surface.

TACK WELDING THE PIECES

Tack welding aluminium is an essential step to prevent distortions due to its high thermal conductivity.

- ✓ **Alignment:** Position the pieces at 90° and use a metal square to verify accuracy. Correct alignment before tacking is crucial.
- ✓ **Tack welds:** Place three to four tacks at the ends and centre of the joint (on the rear side). This helps prevent distortion due to heat expansion. Aluminium expands and deforms easily, so tack welds should be small yet strong.
- ✓ **Verification:** After tack welding, recheck the alignment of the pieces.

FILLET WELD EXECUTION

Once tack welding is complete, you can proceed with the fillet weld. Here, current ramps and pulsed arc parameters will be applied.

PULSED ARC WELDING PARAMETERS

- **Voltage:** 16–18 volts to ensure proper fusion without overheating the material.
- **Pulse frequency:** Higher frequency provides better control of the weld pool, helping to avoid thermal overload in aluminium.
- **Inductance:** A value of +1 or +2 (depending on the equipment) is suitable for keeping the arc stable.
- **Arc length:** Maintain a stick-out of 0.2–0.4″ (5–10mm) between the nozzle and the material to ensure consistent gas protection and prevent oxygen inclusions or porosity.

APPLICATION OF CURRENT RAMPS

1. **Initial ramp-up:**
   - ✓ Set the equipment to apply a current ramp 50% higher than the peak current for 2 seconds.
   - ✓ This allows the material to preheat smoothly before switching to pulsed current, which is critical to prevent cracks or initial deformations in aluminium.
2. **Welding phase:**
   - ✓ Use pulsed mode to improve heat control.

✓ Aluminium has high thermal conductivity, so pulsed arc helps regulate heat input and minimises distortions.

✓ In MIG aluminium welding, always use the push technique (torch oriented forwards) to optimise gas protection and reduce oxidation in the weld area.

3. **Final ramp-down:**

✓ When finishing the weld bead, apply a ramp-down reducing current to 30% of peak value for 3 seconds.

✓ This gradual cooling prevents solidification cracking and ensures uniform solidification of the weld bead.

TORCH MOVEMENT

✓ **Straight motion:** Maintain a steady, straight movement or a slight forward-backward weave, with a torch inclination of approximately 45°. Keep the arc on the edge of the joint to ensure good fusion of both pieces.

✓ **Travel speed control:** Maintain a consistent travel speed to avoid overheating the piece while ensuring sufficient fusion for a homogeneous, porosity-free weld bead (Figure 3.34).

MIG WELDING WITH 5086 ALUMINIUM

PRACTICE 22. BUTT JOINT IN HORIZONTAL POSITION. PA (1G)

- **Base material:** 5086 aluminium sheet, dimensions: 5.91″ x 1.57″ x 0.08″ (150 x 40 x 2 mm).
- **Wire diameter and designation:** 0.047″ (1.2mm) wire, ER 5356 or 4043 (AWS A5.10).
- **Number of beads:** One bead with a straight movement, with the torch oriented forwards (pushing) to achieve better gas shielding and prevent oxidation in the weld.
- **Welding voltage:** 15–17 volts
- **Wire feed speed root pass:** 10–16 ft/min (3–5 m/min).
- **Stick-out length:** 0.2–0.4″ (5–10mm) for short-circuit arc, 0.4–0.6″ (10–15mm) for pulsed arc.
- **Shielding gas flow rate:** 21–32 cubic feet per hour (10–15 L/min) of pure argon, 1 litre per millimetre of nozzle inner diameter ≈ 0.06 ft³ per inch.
- **Transfer mode:** Short-circuit or pulsed arc.

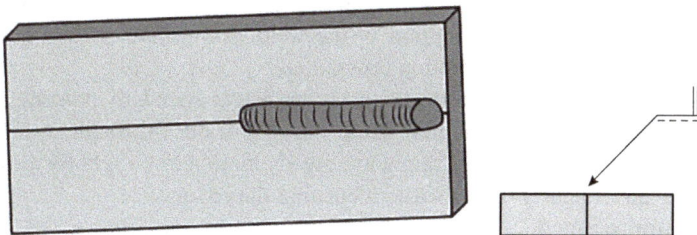

FIGURE 3.34 Technical illustration of a butt joint with a single bevel preparation, showing a weld bead applied at the root.

1. **Material preparation**
   - ✓ **Cleaning:** Aluminium sheets get dirty and oxidise quickly, so it is essential to clean both sides of the sheet using a suitable degreaser or acetone.
   - ✓ **Brushing:** Use a stainless steel wire brush dedicated to aluminium to remove any aluminium oxide (alumina) before welding.
   - ✓ **No gap required:** Since this is a full-butt joint with no gap, no root opening is necessary. Aluminium conducts heat efficiently, so it can melt and fuse properly without a gap, as long as heat input is well controlled.

2. **Tack welding the pieces**
   - ✓ **Alignment:** Ensure that the sheets are perfectly aligned with no mismatch at the edges. Aluminium warps quickly if heat is not controlled, so alignment must be precise.
   - ✓ **Tack welds:** Apply alternating tack welds at both ends and in the centre to prevent distortion during welding. Aluminium expands significantly when heated, so small but strong tack welds will help maintain material shape.

3. **Welding parameters**
   Recommended pulsed arc parameters (1.2 mm ER 5356 Wire):
   - ✓ **Voltage:** 15–17 volts. This stabilises the arc without excessive heat input, preventing sheet deformation.
   - ✓ **Pulse frequency:** A high pulse frequency allows precise weld pool control, essential for thin materials (0.08″ or 2mm) and reduces thermal distortion.
   - ✓ **Inductance:** A medium value of +1 or +2 helps maintain a smooth arc, preventing overcooling, which could lead to porosity or lack of fusion.
   - ✓ **Arc height:** Keep the arc 0.3–0.5″ (8–12mm) high to ensure proper gas shielding and stable fusion. A greater distance can cause lack of fusion, while a shorter distance may lead to burn-through at the edges.

CURRENT RAMP SETTINGS

- **Initial ramp-up:**
  - ✓ 50% of base current for 1–2 seconds. This gently preheats the material, reducing the risk of distortion or cracking.
  - ✓ This gradual increase in heat prevents sudden heat input, which could damage thin aluminium sheets.
- **Final ramp-down:**
  - ✓ 50% of base current for 2–3 seconds. This allows gradual cooling, preventing cracks due to rapid cooling or shrinkage.

4. **Welding process**
   - ✓ **Push technique:** Use a forward torch angle (pushing technique) to improve gas coverage and minimise oxidation in the weld. Aluminium is highly reactive, so adequate shielding gas protection is essential.
   - ✓ **Travel speed:** Maintain a steady and fast travel speed to prevent overheating. Pausing too long in one area may cause warping or burn-through.
   - ✓ **Straight movement:** Use a straight and steady motion or a slight forward-backward weave to ensure good fusion without burning the edges.

5. **Root bead appearance**

The root bead on the back side should meet the following criteria:

✓ **Surface appearance:** The root should be smooth, with no visible porosity or inclusions. In aluminium, the weld will have a shiny or matte appearance, depending

on the shielding gas used, but should not show dark oxidation stains (black or grey discolouration).
- ✓ **Root reinforcement height:** The root reinforcement should be 0.04–0.08″ (1–2mm) above the surface, indicating good penetration without excess material build-up.
- ✓ **Root bead width:** The width of the root bead should match the weld joint width. A narrow bead indicates lack of fusion, while a wide bead may suggest excess heat input or improper gas shielding.

6. **Common defects and contamination indicators**
   - ✓ **Porosity:** Small gas bubbles in the weld indicate porosity, often caused by moisture, oil, or atmospheric gas contamination in the weld pool.
   - ✓ **Oxidation:** Dark or blackened root welds suggest oxidation due to inadequate gas shielding. This weakens corrosion resistance and may lead to future cracks or fractures.
   - ✓ **Lack of penetration:** If the root is not visible or barely raised, it means the weld has not fused properly to the back side, creating a weak and unreliable joint.
   - ✓ **Possible causes:** Low current settings or excessively high travel speed.

7. **Alumina: A Natural Protective Barrier**
   Aluminium naturally forms an aluminium oxide layer (alumina) on its surface, which acts as a protective barrier against the atmosphere.
   - ✓ Alumina is extremely strong, with a melting point of nearly 3,600°F (2,000°C), much higher than aluminium's melting point of 1,220°F (660°C).
   - ✓ Think of alumina as an "airtight seal" around molten aluminium, preventing oxygen and other gases from contaminating the weld pool.
   - ✓ However, during welding, this barrier must be removed to allow proper fusion of the underlying material.
   - ✓ In TIG and MIG welding, the shielding gas (argon or argon/helium mix) acts as a protective shield, preventing new oxide formation while the weld is being made.
   - ✓ If gas coverage is inadequate, oxygen reacts instantly with molten aluminium, forming oxides and weld defects.

CONCLUSION

- ✓ Welding thin aluminium sheets without a root gap requires precise parameter control and heat management.
- ✓ Pulsed arc welding, combined with current ramps, helps distribute heat evenly, reducing distortion and cracking risks.

"Only metal that embraces the fire's heat can transform into a masterpiece; just as only those who accept change can discover their true strength."

Challenges and effort not only shape our skills but also our identity, elevating us to a higher level of understanding and mastery.

INSPECTION PROTOCOL FOR CSWIP WELDING INSPECTORS— BUTT AND FILLET JOINTS USING MIG ON 5086 ALUMINIUM

To inspect and verify that MIG welds performed on 5086 aluminium test coupons using 5356 filler wire (3mm for fillet welds and 2mm for butt welds) have been properly executed in compliance with **ISO 9606–2**, the **CSWIP welding inspector** must follow **a rigorous inspection**

protocol. This process involves visual inspections as well as destructive and non-destructive testing to ensure that the welds meet the required quality standards.

Initial visual inspection

For both 3mm fillet welds and 2mm butt welds:

✓ **Overall weld appearance:** The weld bead must be uniform, with no spatter, undercuts, burn-through, or lack of fusion. The weld must be free from visible porosity or cracks.
✓ **Bead dimensions:** The width and height of the weld bead must conform to the WPS (Welding Procedure Specification) and be consistent throughout the joint.
✓ **Root bead (for butt welds):** The root bead on the back side must show adequate penetration, with no excessive reinforcement or undercut. A uniform reinforcement and width are indicators of good fusion.
✓ **Oxidation and thermal discolouration:** Since aluminium is highly reactive, shielding gas coverage must be sufficient to prevent oxidation or excessive heat discolouration, which may indicate exposure to atmospheric contaminants.

Non-destructive testing (NDT)

To complement visual inspection, the following non-destructive tests (NDT) are applied:

✓ **Liquid penetrant testing (PT):**
• Applied to both coupons, especially the butt joint, to detect surface cracks, porosity, or hidden defects.
• Penetrant testing is highly effective in identifying defects that may compromise weld strength.
✓ **Industrial radiography (RT):**
• Radiographic testing may be used, particularly on butt welds, to detect internal porosity, inclusions, or lack of fusion that may not be externally visible.
• X-ray inspection is highly useful for aluminium welds, ensuring that no internal defects affect weld integrity.

Destructive testing

Guided bend test (fracture testing)

This test is critical for welder qualification under ISO 9606–2 and applies to both types of test coupons. It evaluates the ductility and strength of the weld under deformation.

✓ **For 3mm fillet welds:**
• A section of the weld bead is selected for a bend test to check ductility.
• The sample is bent over a mandrel up to a specified angle.
• If the weld shows cracks, excessive porosity, or defects in the weld bead or root, they will appear during fracture testing.
✓ **For 2mm butt welds:**
• Face and root bend tests are performed, where the coupon is bent on both the weld face and root side.

- This test verifies that no internal defects (such as cracks or lack of fusion) are present.
- For butt welds, root fusion is crucial for joint strength, making this test particularly critical.

PENETRATION MEASUREMENT

✓ For butt welds, it is essential to measure the penetration depth of the root bead.
✓ This can be done using a vernier calliper or micrometre to confirm compliance with WPS requirements.
✓ For a 2mm weld, complete penetration with minimal reinforcement is required.

ADDITIONAL TESTING (IF REQUIRED)

Depending on project specifications or client requirements, additional tests may be conducted:

✓ **Ultrasonic testing (UT):**
  - Less common for thin materials, but occasionally used to verify weld integrity without destroying the coupon.
✓ **Micrographic Examination:**
  - In some cases, metallographic analysis may be performed to examine weld microstructure and the HAZ.
  - This is useful for verifying whether welding temperature and cooling rates were properly controlled.

FINAL EVALUATION AND REPORT

Once all inspections and tests are completed, the CSWIP welding inspector prepares a detailed report including:

✓ Results of visual and non-destructive inspections.
✓ Description of any detected defects.
✓ Results of destructive testing (bend tests, etc.).
✓ Final conclusions on the compliance of the coupons with ISO 9606–2 standards.

This inspection protocol ensures that the welds have been executed correctly and meet the specified quality and safety requirements.

If you wish to practise this exercise with a qualification test coupon, the minimum dimensions (in millimetres) according to UNE EN ISO 9606–2 are as follows:

INTRODUCTION TO FLUX-CORED WIRE WELDING: HISTORY, TYPES, CHARACTERISTICS, AND APPLICATIONS

Flux-cored wire welding has evolved significantly since its introduction to the industry in the 1950s. The development of this process was driven by the need for more efficient and versatile welding methods, allowing faster deposition rates and higher quality welds under varied working conditions. Unlike solid wires, flux-cored wires feature a hollow core, which can be filled

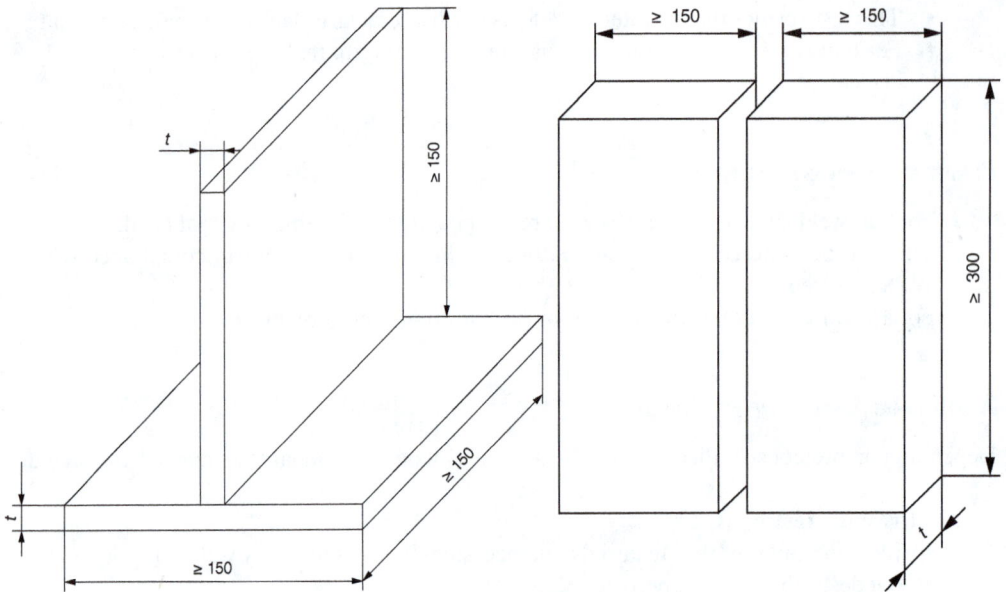

FIGURE 3.35 Technical illustrations showing the dimensions of welding test coupons. The first image depicts a butt joint between two rectangular plates with a bevel, while the second represents a fillet weld joint with a vertical plate on a horizontal base. Both images include minimum dimensions and relevant geometric details.

with deoxidisers, slag-forming agents, and other chemical elements. This composition provides distinct advantages in different working environments, expanding their range of applications (Figure 3.35).

Initially, flux-cored wires were widely used for welding thick and high-strength materials in heavy industries, including shipbuilding, petrochemical plants, and steel structures. Their ability to handle greater material thicknesses and facilitate welding in difficult positions made them a preferred option for projects where efficiency and weld quality were critical factors. Today, advances in wire design and composition have further expanded their applications, making them suitable for a broad range of industrial sectors.

TYPES OF FLUX-CORED WIRES

There are several types of flux-cored wires, each designed to meet specific needs. The following section covers the main types and their key characteristics:

METAL-CORED WIRE

This type of wire combines a metallic core with a tubular structure, improving fusion and weld metal quality. Metal-cored wires are ideal for applications requiring high productivity and excellent mechanical properties, such as automotive manufacturing and heavy structural fabrication.

✓ **Key advantages:**
- Provides **deep and uniform penetration**, particularly in **thick materials**.
- Produces **less fume and spatter** compared to other flux-cored wires.
- Ideal for **high-speed** and **automated welding processes**.

RUTILE FLUX-CORED WIRE

This type of wire contains rutile-based components in its core, improving arc stability and facilitating slag removal. It is commonly used in applications requiring aesthetically pleasing welds and easy slag removal.

- ✓ **Key advantages:**
  - Provides a **stable welding arc**.
  - Produces **smooth, visually appealing welds**.
  - **Easy post-weld cleaning**, with minimal slag residue.
- ✓ **Main limitation:**
  - Lower penetration than other flux-cored wires, making it more suitable for medium-thickness and thin materials.

BASIC FLUX-CORED WIRE

Basic flux-cored wires contain powerful deoxidisers, enhancing mechanical strength and crack resistance. These wires are well-suited for structurally demanding applications and harsh environmental conditions.

- ✓ **Key advantages:**
  - Excellent **mechanical properties**.
  - High **tolerance to surface contamination**.
  - **Superior impact toughness**, even at **low temperatures**.
- ✓ **Main limitation:**
  - Requires **higher operator skill levels** due to **increased slag formation**.
  - **Slower welding speeds** compared to rutile wires.

SELF-SHIELDED FLUX-CORED WIRE

As the name suggests, self-shielded wires do not require an external shielding gas, as the core generates a protective gas atmosphere upon burning. This makes them ideal for outdoor applications and hard-to-reach locations.

- ✓ **Key advantages:**
  - No need for **shielding gas**, making them highly **portable**.
  - Well-suited for **construction, infrastructure repairs, and field welding**.
- ✓ **Main limitations:**
  - Produces **higher levels of smoke and spatter**.
  - Weld appearance is **less aesthetic** than gas-shielded wires.

GENERAL CHARACTERISTICS OF FLUX-CORED WIRES

Flux-cored wires are valued for their versatility. Their hollow structure allows for chemical additives that improve welding performance. Some key characteristics to consider include:

- ✓ **Penetration and weld pool control:**
  - Metal-cored and basic flux-cored wires provide deep penetration, making them ideal for thick joints and out-of-position welding.

✓ **Reduced defects and arc stability:**
  - The chemical components inside these wires help reduce porosity and arc instability, improving weld quality across different positions.
✓ **Fume and slag generation:**
  - Rutile, basic, and self-shielded flux-cored wires produce slag, requiring cleaning between passes.
  - Self-shielded wires generate higher fume emissions, making proper ventilation essential.

APPLICATIONS OF FLUX-CORED WIRES

Thanks to their diverse characteristics, flux-cored wires are used across a wide range of industries:

✓ **Shipbuilding and heavy infrastructure**
  - Basic and metal-cored wires are preferred due to their high penetration and mechanical strength.
  - Used for large-scale structural fabrication requiring high durability and reliability.
✓ **Automotive industry and machinery fabrication**
  - Metal-cored wires are commonly used for high-speed production welding.
  - Their low slag formation makes them ideal for precision welding applications.
✓ **Construction and outdoor repairs**
  - Self-shielded wires are widely used in building construction and infrastructure repairs.
  - No need for shielding gas makes them ideal for fieldwork and remote locations.
✓ **General fabrication and workshop applications**
  - Rutile flux-cored wires are commonly used in workshops and general manufacturing.
  - They offer ease of use, quick cleaning, and smooth weld aesthetics.

CONCLUSION

The selection of flux-cored wire depends on the specific application and working conditions. Each wire type has strengths and limitations, making it better suited for certain environments and objectives. Over time, flux-cored wires have proven to be an essential tool in modern welding, enabling welders to tackle both precision tasks in workshops and large-scale structural projects in the field.

By understanding their properties, welders can maximise performance and achieve high-quality welds tailored to the specific demands of each project.

FLUX-CORED WIRE WELDING

PRACTICE 23. HORIZONTAL FILLET WELD WITH METAL-CORED WIRE—PB(2F)

- **Base material:** Carbon steel plate 5.91″ x 1.57″ x 0.31″ (150 x 40 x 8 mm)
- **Wire diameter and designation:** 0.047″ (1.2mm) Böhler HL 51 T MC
- **Number of beads:** Three beads, all applied using a straight movement
- **Welding voltage:** 23–28 volts *(refer to the wire data sheet for precise settings)*
- **Wire feed speed:** 20–39 ft/min (6–12 m/min)
- **Stick-out length:** 0.2–0.4″ (5–10mm)

- **Shielding gas flow rate:** 32 cubic feet per hour (15 L/min) – argon (85%)/CO_2 (20%) mixture
- **Transfer mode:** Spray transfer

INTRODUCTION

The technique used in this practice is very similar to that of Practice 7 (horizontal fillet weld using solid MAG wire). However, this section outlines the advantages and limitations of metal-cored flux-cored wire compared to solid wire.

ADVANTAGES

IMPROVED ARC STABILITY

Metal-cored wires provide a more stable and controlled arc than solid wires, resulting in a more uniform and controlled weld pool.

GREATER PENETRATION AND UNIFORMITY

Metal-cored wires allow for deeper and more consistent penetration in carbon steel, making them ideal for 8mm thick joints.

INCREASED PRODUCTIVITY

Compared to solid wires, metal-cored wires enable higher welding speeds, reducing overall welding time and improving efficiency.

REDUCED SPATTER

The process generates less spatter, significantly reducing post-weld cleaning time.

EXCELLENT WELD METAL CLEANLINESS

Minimal slag formation makes slag removal quick and effortless, leaving behind a clean and smooth weld bead.

LOWER SENSITIVITY TO SURFACE CONTAMINATION

Metal-cored wires are less affected by surface contaminants such as light oils and rust, making them suitable for less-than-ideal surface preparation conditions.

VERSATILITY IN WELDING POSITIONS

While particularly effective in horizontal and flat positions, metal-cored wires can be used in various welding positions with the appropriate technique.

LIMITATIONS

HIGHER FUME EMISSIONS

Metal-cored wires produce more welding fumes than solid wires, requiring adequate ventilation or fume extraction systems to ensure operator safety.

HIGHER CONSUMABLE COSTS

Metal-cored wires are more expensive than solid wires, increasing overall welding costs.

REQUIRES SUITABLE EQUIPMENT

A welding machine capable of spray transfer mode is essential, preferably with synergic programme settings to optimise welding parameters.

INCREASED SHIELDING GAS CONSUMPTION

Gas consumption is higher than with solid wires due to:

- Spray transfer mode, which requires a consistent gas flow.
- The wire's specific formulation, which influences gas flow requirements.

MORE DEMANDING WELDING TECHNIQUE

Although both push and drag techniques can be used, dragging (pulling) is generally more effective in this case. However, this technique may require more skill, especially for less experienced welders.

POSSIBLE SLAG BUILD-UP IN NON-IDEAL POSITIONS

Although slag formation is minimal, in certain positions, slag accumulation may occur, requiring cleaning between passes.

RISK OF POROSITY IF PARAMETERS ARE NOT PROPERLY ADJUSTED

Metal-cored wire can be prone to porosity if:

- Wire feed speed or current is inconsistent (causing an unstable arc).
- Surface preparation is inadequate.
- Shielding gas flow is too low or too high.
- Stick-out length is excessive, affecting arc stability.

FLUX-CORED WIRE WELDING

PRACTICE 24. HORIZONTAL FILLET WELD WITH RUTILE FLUX-CORED WIRE—PB(2F)

- **Base material:** Carbon steel plate 5.91" x 1.57" x 0.31" (150 x 40 x 8 mm)
- **Wire diameter and designation:** 0.047" (1.2mm) Böhler Ti 52 T FD (HP)

- **Number of beads:** Three beads, all applied using a straight movement
- **Welding voltage:** 23–28 volts *(refer to the wire data sheet for precise settings)*
- **Wire feed speed:** 20–39 ft/min (6–12 m/min)
- **Stick-out length:** 0.2–0.4″ (5–10mm)
- **Shielding gas flow rate:** 32 cubic feet per hour (15 L/min) – argon (85%)/CO_2 (20%) mixture
- **Transfer mode:** Spray transfer

INTRODUCTION

This practice describes the advantages and limitations of rutile flux-cored wire welding compared to solid wire, metal-cored wire, and basic flux-cored wire when welding the same type of joint.

ADVANTAGES

1. **Easy to use and control**
   - Rutile flux-cored wire provides a smooth operation that is easy to control, especially in a horizontal position.
   - Compared to solid wires and basic flux-cored wires, it offers a more stable and forgiving arc, making it ideal for less experienced welders.
2. **Good weld bead appearance**
   - Produces clean and smooth weld beads, requiring less post-weld finishing compared to basic flux-cored wires, which tend to produce more spatter and slag.
3. **Better tolerance to contaminated surfaces**
   - More tolerant of light surface contaminants, such as minor oil residues or oxidation, compared to solid wire, which requires stricter surface preparation.
4. **Easy-to-remove slag**
   - Rutile slag is easier to remove than that of basic flux-cored wire, simplifying post-weld cleaning and increasing efficiency.
5. **Lower shielding gas cost**
   - While shielding gas is required, its gas consumption is less critical.
   - Can sometimes be used with standard CO_2/argon mixtures, reducing costs compared to metal-cored wires, which often require more specific gas mixtures.
6. Suitable for conventional welding equipment
   - Can be used with standard welding machines, without requiring advanced synergic programs.
   - Compatible with a wider range of welding machines, making it easier to implement across different work environments.

LIMITATIONS

1. Higher fume emissions
   - Like all flux-cored wires, it produces more fumes than solid or metal-cored wires.
   - Proper ventilation or fume extraction systems are required to protect the welder's health.
2. **Moderate penetration**
   - Penetration is shallower and less uniform compared to metal-cored or basic flux-cored wires.
   - Not ideal for applications requiring full penetration on thicker materials.

3. **Moderate welding speed**
   - Welding speed is acceptable, but not as high as with metal-cored wires.
   - May be less efficient in high-productivity applications where faster travel speeds are essential.

4. **More sensitive to welding position**
   - Performs very well in horizontal and flat positions.
   - Less effective in vertical-up or overhead welding compared to basic flux-cored wire, which performs better in out-of-position welds.

5. **Higher consumable costs**
   - More expensive than solid wire, although cheaper than metal-cored wire.
   - May increase operational costs compared to solid wire in applications where solid wire is viable.

6. **Risk of porosity in adverse conditions**
   - While it tolerates some surface contamination, it is susceptible to porosity if:
     - o Surface cleaning is inadequate.
     - o Gas shielding is not optimised.
     - o Stick-out length is excessive.
     - o Humidity levels are high.

7. Recommended use of dragging technique
   - The dragging (pulling) technique is recommended for better slag control and weld stability.
   - Less intuitive for inexperienced welders, requiring additional skill and control.

FLUX-CORED WIRE WELDING

PRACTICE 25. HORIZONTAL FILLET WELD WITH BASIC FLUX-CORED WIRE—PB(2F)

- **Base material:** Carbon steel plate 5.91" x 1.57" x 0.31" (150 x 40 x 8 mm)
- **Wire diameter and designation:** 0.047" (1.2mm) Böhler KB 46 T FD
- **Number of beads:** Three beads, all applied using a straight movement
- **Welding voltage:** 23–28 volts *(refer to the wire data sheet for precise settings)*
- **Wire feed speed:** 20–39 ft/min (6–12 m/min)
- **Stick-out length:** 0.2–0.4" (5–10mm)
- **Shielding gas flow rate:** 32 cubic feet per hour (15 L/min) – argon (85%)/CO_2 (20%) mixture
- **Transfer mode:** Spray transfer

INTRODUCTION

This practice describes the advantages and limitations of basic flux-cored wire welding compared to solid wire, metal-cored wire, and rutile flux-cored wire when welding the same type of joint.

ADVANTAGES

1. **Excellent penetration and strength**
   - Basic flux-cored wire provides deep and consistent penetration, making it ideal for thicker materials, such as this 8mm steel plate.
   - The resulting welds are highly resistant and reliable, making them suitable for structural applications.

2. **High metallurgical cleanliness**
   - Produces weld beads with superior metallurgical cleanliness, reducing impurities and minimising trapped oxygen.
   - This leads to higher-quality welds compared to rutile flux-cored and metal-cored wires.
3. **Good arc stability in challenging conditions**
   - Maintains arc stability even in difficult welding conditions, such as out-of-position welding.
   - While this is a horizontal weld, basic flux-cored wire performs well in a variety of welding positions.
4. **Superior performance in high-strength applications**
   - Due to its high-strength deposit and improved cleanliness, this wire is preferred for applications requiring high mechanical resistance, such as in construction and structural welding.
   - It is a better choice than rutile wire, which is less suitable for high-strength requirements.
5. Protective slag formation
   - The slag layer shields the weld pool from atmospheric contamination.
   - Compared to rutile flux-cored wire, the slag is more controlled, providing enhanced protection during the welding process.

LIMITATIONS

1. **Increased slag formation and removal effort**
   - The slag formed by basic flux-cored wire is thicker and harder to remove compared to rutile or metal-cored wires.
   - Requires more cleaning time, which can reduce productivity.
2. **Higher fume emissions**
   - Produces significantly more fumes and gases than solid wire and metal-cored wire.
   - Proper ventilation or fume extraction systems are essential, particularly in enclosed workspaces.
3. **More sensitive to surface contamination**
   - Unlike rutile and metal-cored wires, basic flux-cored wire is more sensitive to surface contamination, such as oil and oxides.
   - Proper surface preparation is crucial to prevent welding defects, adding extra steps before welding.
4. **Requires dragging technique (torch facing backwards)**
   - For better slag control and weld stability, the dragging technique (pulling the torch backwards) is recommended.
   - This may require greater skill, especially in more challenging welding positions.
5. **Higher consumable cost**
   - Basic flux-cored wire is more expensive than rutile flux-cored and solid wires.
   - This increases production costs compared to these alternatives.
6. **More demanding welding technique**
   - Welding with basic flux-cored wire can be more complex, requiring better control of the slag and bead stability.
   - Less suitable for beginners, as it may require additional training.
7. **Moderate welding speed**
   - Although it ensures good penetration and arc stability, welding speed is lower compared to metal-cored wire.
   - This can be a drawback in high-productivity applications.

SELF-SHIELDED FLUX-CORED WIRE WELDING

PRACTICE 26. HORIZONTAL FILLET WELD WITH SELF-SHIELDED FLUX-CORED WIRE—PB(2F)

- **Base material:** Carbon steel plate 5.91″ x 1.57″ x 0.31″ (150 x 40 x 8 mm)
- **Wire diameter and designation:** 0.047″ (1.2mm) Böhler TI 52 NG T FD
- **Number of beads:** Three beads, all applied using a straight movement
- **Welding voltage:** 23–28 volts *(refer to the wire data sheet for precise settings)*
- **Wire feed speed:** 20–39 ft/min (6–12 m/min)
- **Stick-out length:** 0.2–0.4″ (5–10mm)
- **Shielding gas flow rate:** Not required (self-shielded wire)
- **Transfer mode:** Open-arc transfer (spray transfer does not apply)
- **Polarity**: May require reverse polarity (torch connected to negative).

INTRODUCTION

This practice describes the advantages and limitations of self-shielded flux-cored wire welding, compared to solid wire, metal-cored wire, and rutile or basic flux-cored wires when welding the same type of joint.

ADVANTAGES

1. **No need for shielding gas**
   - Since this is a self-shielded wire, it does not require external shielding gas, making it ideal for outdoor applications or windy environments, where shielding gas can disperse easily.
   - This is a major advantage over solid wire, metal-cored, and gas-shielded flux-cored wires, which require shielding gas.
2. **Versatility in outdoor environments**
   - Self-shielded wire is particularly useful in environments where atmospheric conditions cannot be controlled, such as outdoor welding or hard-to-reach areas.
   - This makes it a preferred choice for field applications and structural maintenance work.
3. **Portable and lightweight equipment**
   - No gas cylinders required, meaning the equipment is lighter and easier to transport.
   - This offers a significant advantage over solid wire, metal-cored, and gas-shielded flux-cored wires, which require bulky gas bottles.
4. **Adequate penetration for medium thicknesses**
   - Provides sufficient penetration on thicknesses up to 0.31″ (8mm), similar to rutile flux-cored wire.
   - However, penetration is generally lower compared to basic flux-cored wire.
5. **Lower operating costs**
   - No shielding gas expenses, reducing operational costs compared to gas-shielded flux-cored wires, solid wires, and metal-cored wires.
6. **Moderate tolerance to slightly oxidised surfaces**
   - Offers some tolerance to light surface oxidation, allowing welding without extensive material preparation.
   - However, not as tolerant as rutile flux-cored wire.

LIMITATIONS

1. **Increased fume emissions and toxic gases**
   - Produces significantly more fumes and gases than solid wire and metal-cored wire.
   - Adequate ventilation or respiratory protection is necessary, especially in confined spaces.
2. **Lower weld quality and less aesthetic appearance**
   - Compared to rutile flux-cored wire and solid wire, weld beads tend to be rougher and less uniform.
   - May require additional post-weld finishing and cleaning in applications where appearance is important.
3. **Prone to slag-related defects**
   - Slag formation is more difficult to control and remove, compared to rutile flux-cored wires.
   - Increases cleaning time, affecting productivity compared to rutile and metal-cored wires.
4. **Lower penetration than basic flux-cored wire**
   - While penetration is acceptable, it is less than that of basic flux-cored wire.
   - This makes it less suitable for high-strength welding applications requiring deep fusion on thick materials.
5. **Limited positional welding capability**
   - Works well in flat and horizontal positions, but can be difficult to control in vertical and overhead positions.
   - Performs worse in positional welding compared to rutile and basic flux-cored wires.
6. **Slower welding speed**
   - Welding speed is generally slower compared to metal-cored wire, reducing productivity.
   - This is especially relevant in high-production environments.
7. **Requires dragging technique for better control**
   - Similar to basic flux-cored wire, self-shielded wire performs best when dragged, rather than when pushed.
   - This requires more skill from the welder to maintain a stable bead and avoid defects.

METAL-CORED WIRE WELDING

PRACTICE 27. HORIZONTAL BUTT JOINT WITH BEVELLED PLATES USING METAL-CORED WIRE—PA(1G)

✓ **Base material:** Carbon steel plate 5.91″ x 1.57″ x 0.31″ (150 x 40 x 8 mm)
✓ **Wire diameter and designation:** 0.047″ (1.2mm) Böhler HL 51 T MC
✓ **Number of beads:**
   - **Root pass:** Straight movement, drag technique (torch oriented backwards)
   - **Fill and cap passes**: Lateral movement
✓ **Welding voltage:**
   - **Root pass:** 16–20 volts
   - **Fill and cap passes:** 18–22 volts (refer to the wire data sheet for precise settings.
✓ **Wire feed speed:**
   - **Root pass:** 10–16 ft/min (3–5 m/min)
   - **Remaining passes:** 13–30 ft/min (4–9 m/min)
✓ **Stick-out length:** 0.2–0.4″ (5–10mm) for short-circuit arc
✓ **Shielding gas flow rate:** 32 cubic feet per hour (15 L/min) of 80% argon/20% CO_2

✓ **Transfer mode:** Short-circuit
✓ **Bevel angle:** 35° V preparation
✓ **Additional considerations:** 1–2 mm root face, 2.5 mm root gap

COMPARISON WITH OTHER WIRES FOR BEVELLED BUTT JOINTS

1. Advantages over solid wire
   - ☑ **Fewer passes required:** Due to its higher penetration and arc stability, metal-cored wire may require fewer passes than solid wire, reducing work time and effort for thick butt joints.
   - ☑ **Greater tolerance to parameter variations:** Metal-cored wire adapts better to fluctuations in wire feed speed and voltage, whereas solid wire is more sensitive to parameter changes, which could affect weld quality.
2. **Advantages over rutile flux-cored wire**
   - ☑ **Superior root fusion:** Metal-cored wire provides better fusion at the root compared to rutile flux-cored wire, which is crucial for butt welds on thick steel to ensure uniform penetration from the first pass.
   - ☑ **Reduced slag inclusions:** Minimal slag formation helps prevent inclusions in the final weld, improving joint integrity compared to rutile flux-cored wire.
3. **Advantages over basic flux-cored wire**
   - ☑ **Lower weld hardness:** Metal-cored welds produce a softer bead, making them more ductile and less prone to brittle fractures compared to basic flux-cored wire, which has higher hardness levels.
   - ☑ **No preheating required:** Unlike basic flux-cored wire, which often requires preheating to prevent cracking in certain carbon steels, metal-cored wire allows welding without preheating, saving time and resources.
4. **Advantages over self-shielded wire**
   - ☑ **Better suited for workshop environments:** While self-shielded wire is ideal for outdoor work, metal-cored wire excels in workshop conditions, offering a more stable arc and smoother bead appearance, improving quality control and inspection in butt welds.
   - ☑ **Lower risk of porosity:** Metal-cored wire produces fewer porosity-related defects compared to self-shielded wire, making it more reliable in high-demand applications where dense, defect-free welds are required.

DISADVANTAGES

1. **Disadvantages compared to solid wire**
   - ⚠ **Requires precise parameter adjustment:** Metal-cored wire is more sensitive to parameter settings, whereas solid wire is more forgiving, making it easier for less experienced operators.
   - ⚠ **Higher sensitivity to travel speed variations:** In butt joints, any inconsistency in travel speed can significantly impact fusion when using metal-cored wire, requiring more skill compared to solid wire.
2. **Disadvantages compared to rutile flux-cored wire**
   - ⚠ **Limited usability in out-of-position welding:** While metal-cored wire performs well in horizontal and flat positions, rutile flux-cored wire is superior in vertical and overhead welding.
   - ⚠ **Less cost-effective for thin materials:** For thin material applications, rutile flux-cored wire can be more economical without the additional costs of metal-cored wire consumables.

3. **Disadvantages compared to basic flux-cored wire**
   ⚠ **Lower tensile strength:** Metal-cored welds generally do not achieve the same high tensile strength as basic flux-cored wire, making the latter better suited for critical structural joints.
   ⚠ **Requires cleaner surfaces:** Basic flux-cored wire tolerates surface contaminants better, while metal-cored wire requires thorough cleaning to prevent weld defects, increasing preparation time.
4. **Disadvantages compared to self-shielded wire**
   ⚠ **Always requires shielding gas:** Unlike self-shielded wire, metal-cored wire always requires gas, limiting its outdoor use and increasing operating costs.
   ⚠ **Less adaptable to outdoor and windy conditions:** Metal-cored wire is less effective than self-shielded wire in outdoor environments, where shielding gas can become unstable.

FLUX-CORED WIRE WELDING

PRACTICE 28. HORIZONTAL BUTT JOINT WITH BEVELLED PLATES USING RUTILE FLUX-CORED WIRE—PA(1G)

✓ **Base material:** Carbon steel plate 5.91″ x 1.57″ x 0.31″ (150 x 40 x 8 mm)
✓ **Wire diameter and designation:** 0.047″ (1.2mm) Böhler Ti 52 T FD (HP)
✓ **Number of beads:**
   • **Root pass:** Straight movement, drag technique (torch oriented backwards)
   • **Fill and cap passes:** Lateral movement
✓ **Welding voltage:**
   • **Root pass:** 16–20 volts
   • **Fill and cap passes:** 18–22 volts (refer to the wire data sheet for precise settings
✓ **Wire feed speed:**
   • **Root pass:** 10–16 ft/min (3–5 m/min)
   • **Remaining passes:** 13–30 ft/min (4–9 m/min)
✓ **Stick-out length:** 0.2–0.4″ (5–10mm) for short-circuit arc
✓ **Shielding gas flow rate:** 32 cubic feet per hour (15 L/min) of 80% argon/20% CO_2
✓ **Transfer mode:** Short-circuit
✓ **Bevel angle:** 35° V preparation
✓ **Additional considerations:** 1–2mm root face, 2.5mm root gap

COMPARISON WITH OTHER WIRES FOR BEVELLED BUTT JOINTS

1. **Advantages over solid wire**
   ☑ **Better weld pool control on thick plates:** Rutile flux-cored wire provides greater weld pool control, making it easier to handle in thicker butt joints, such as this 8mm joint, compared to solid wire, where controlling the weld pool may be more challenging.
   ☑ **Reduced spatter in short-circuit transfer mode:** This type of wire produces less spatter when used with short-circuit transfer, resulting in a cleaner and more efficient welding process.
2. **Advantages over metal-cored wire**
   ☑ **Easier defect correction at the root:** If defects occur at the root, rutile flux-cored wire allows for easier rework and correction, as it fuses well when applying repair welds. In contrast, metal-cored wire can be less flexible for rework.

☑ **Better suited for intermittent welding:** Rutile flux-cored wire provides a more stable arc restart, making it better for intermittent welds, whereas metal-cored wire may experience instability when restarting in short-circuit mode.

3. **Advantages over basic flux-cored wire**

☑ **Faster arc ignition:** Rutile flux-cored wire ignites quickly and reliably, making it ideal for multi-stop butt joints, whereas basic flux-cored wire may have slower and less consistent arc starts.

☑ **Less sensitive to variations in travel speed:** Rutile flux-cored wire tolerates travel speed variations better, allowing more consistent beads even if the operator does not maintain a perfectly steady pace. Basic flux-cored wire, on the other hand, requires greater precision to avoid defects.

4. **Advantages over self-shielded wire**

☑ **Better weld bead appearance:** In butt joints, rutile flux-cored wire produces a smoother, more visually appealing bead than self-shielded wire, which is beneficial for applications where aesthetics matter.

☑ **More stable arc control at varying temperatures:** This wire maintains a more stable arc during long butt joints or fluctuating temperature conditions, which is less manageable with self-shielded wire.

DISADVANTAGES

1. **Disadvantages compared to solid wire**

⚠ **Higher shielding gas consumption:** Rutile flux-cored wire requires a consistent and stable gas flow, leading to higher gas consumption compared to solid wire, which can function effectively with a lower gas flow rate.

⚠ **More sensitive to atmospheric moisture:** This wire is more affected by humidity, which can cause porosity in the weld bead. In contrast, solid wire is less sensitive to moisture.

2. **Disadvantages compared to metal-cored wire**

⚠ **Lower fusion efficiency on bevelled edges:** Although rutile flux-cored wire offers good weld pool control, metal-cored wire fuses better in bevelled joints, achieving a more uniform weld in V-prepared butt joints requiring deep penetration.

⚠ **Limited suitability for high-strength structural applications:** In high-load structural joints, rutile flux-cored wire does not achieve the same mechanical strength as metal-cored wire, limiting its use in critical structural projects.

3. **Disadvantages compared to basic flux-cored wire**

⚠ **Lower tolerance to surface contamination:** Unlike basic flux-cored wire, which handles oxides and contaminants well, rutile flux-cored wire requires cleaner surfaces to avoid defects, meaning additional preparation time.

⚠ **Less effective in out-of-position welding:** While rutile flux-cored wire works well in horizontal positions, its performance declines in out-of-position welding—especially in butt joints, where basic flux-cored wire provides greater stability and control.

4. **Disadvantages compared to self-shielded wire**

⚠ **More setup time due to shielding gas requirements:** Rutile flux-cored wire requires gas shielding, meaning additional setup time compared to self-shielded wire, which can be used immediately in environments without gas supply.

⚠ **Dependent on a controlled environment for defect prevention:** Rutile flux-cored wire is better suited for workshop conditions, where wind and external factors do not affect gas shielding, while self-shielded wire is more versatile for outdoor work.

FLUX-CORED WIRE WELDING

PRACTICE 29. HORIZONTAL BUTT JOINT WITH BEVELLED PLATES USING BASIC FLUX-CORED WIRE—PA(1G)

- ✓ **Base material:** Carbon steel plate 5.91″ x 1.57″ x 0.31″ (150 x 40 x 8 mm)
- ✓ **Wire diameter and designation:** 0.047″ (1.2mm) Böhler KB 46 T FD
- ✓ **Number of beads:**
  - **Root pass:** Straight movement, drag technique (torch oriented backwards)
  - **Fill and cap passes:** Lateral movement
- ✓ **Welding voltage:**
  - **Root pass:** 16–20 volts
  - **Fill and cap passes:** 18–22 volts (refer to the wire data sheet for precise settings)
- ✓ **Wire feed speed:**
  - **Root pass:** 10–16 ft/min (3–5 m/min)
  - **Remaining passes:** 13–30 ft/min (4–9 m/min)
- ✓ **Stick-out length:** 0.2–0.4″ (5–10mm) for short-circuit arc
- ✓ **Shielding gas flow rate:** 32 cubic feet per hour (15 L/min) of 80% argon/20% CO_2
- ✓ **Transfer mode:** Short-circuit
- ✓ **Bevel angle:** 35° V preparation
- ✓ **Additional considerations:** 1–2mm root face, 2.5mm root gap

COMPARISON WITH OTHER WIRES FOR BEVELLED BUTT JOINTS

1. **Advantages over solid wire**
   - ☑ **Higher mechanical strength:** Basic flux-cored wire provides greater tensile strength and hardness in the weld bead compared to solid wire, making it ideal for critical structural applications, such as butt joints that require high load-bearing capacity.
   - ☑ **Excellent fusion on thick materials:** For 8mm or thicker materials, basic flux-cored wire ensures full fusion, especially in the root pass, where solid wire may struggle to penetrate uniformly.
2. **Advantages over metal-cored wire**
   - ☑ **Better tolerance to base metal contamination:** Basic flux-cored wire is less sensitive to surface oxidation and contaminants, which reduces the need for extensive cleaning before welding—useful in projects with limited preparation time.
   - ☑ **Superior multi-position welding capability:** Although this practice is performed in horizontal position (PA/1G), basic flux-cored wire provides better performance in other positions (such as vertical or overhead) compared to metal-cored wire, making it more versatile for complex projects.
3. **Advantages over rutile flux-cored wire**
   - ☑ **Better resistance to high temperatures:** Basic flux-cored wire performs better in high-temperature conditions and is less prone to hot cracking, making it ideal for butt joints that must endure thermal and mechanical loads.
   - ☑ **Stronger impact resistance:** This wire produces a harder weld bead with superior impact resistance, which is advantageous for structural applications where extra mechanical durability is required. In contrast, rutile flux-cored wire tends to be more brittle.

4. **Advantages over self-shielded wire**
   - ☑ **Smoother weld bead finish:** In a controlled environment with shielding gas, basic flux-cored wire creates a more uniform weld bead with fewer porosity issues than self-shielded wire, improving the overall integrity of the joint.
   - ☑ **Less technique-dependent:** Basic flux-cored wire tolerates minor variations in operator technique without compromising weld quality, whereas self-shielded wire requires precise control to prevent defects in critical joints.

DISADVANTAGES

1. **Disadvantages compared to solid wire**
   - ⚠ **Increased slag generation requiring inter-pass cleaning:** Basic flux-cored wire produces more slag than solid wire, requiring additional cleaning between passes to avoid inclusions, which increases welding time.
   - ⚠ **Less suitable for aesthetic welds:** This wire does not produce as smooth a finish as solid wire, which may be undesirable in applications requiring a clean, polished appearance.
2. **Disadvantages compared to metal-cored wire**
   - ⚠ **Higher fume and particle emissions:** Basic flux-cored wire generates more fumes than metal-cored wire, requiring adequate ventilation and respiratory protection, especially in enclosed spaces.
   - ⚠ **More sensitive to parameter variations:** Although it tolerates surface contamination well, basic flux-cored wire requires precise voltage and feed rate adjustments to prevent defects. In contrast, metal-cored wire is more stable with slight parameter fluctuations.
3. **Disadvantages compared to rutile flux-cored wire**
   - ⚠ **More difficult weld pool control in horizontal positions:** While basic flux-cored wire is excellent for structural welding, it can be harder to control in horizontal positions compared to rutile flux-cored wire, which may require greater operator skill.
   - ⚠ **Higher spatter levels:** In short-circuit transfer mode, basic flux-cored wire tends to generate more spatter than rutile flux-cored wire, increasing clean-up time and potentially affecting the weld bead appearance.
4. **Disadvantages compared to self-shielded wire**
   - ⚠ **Requires shielding gas:** Unlike self-shielded wire, basic flux-cored wire requires shielding gas, which increases operational costs and limits its use outdoors, where wind can affect gas stability.
   - ⚠ **Less portable equipment setup:** Since basic flux-cored wire requires a gas supply, the welding setup is bulkier and less portable, making it less practical for remote or hard-to-reach job sites compared to self-shielded wire, which does not require gas.

SELF-SHIELDED FLUX-CORED WIRE WELDING

PRACTICE 30. HORIZONTAL BUTT JOINT WITH BEVELLED PLATES USING SELF-SHIELDED FLUX-CORED WIRE—PA(1G)

- ✓ **Base material:** Carbon steel plate 5.91″ x 1.57″ x 0.31″ (150 x 40 x 8 mm)
- ✓ **Wire diameter and designation:** 0.047″ (1.2mm) Böhler TI 52 NG T FD

✓ **Number of beads:**
  - **Root pass:** Straight movement, drag technique (torch oriented backwards)
  - **Fill and cap passes:** Lateral movement.
✓ **Welding voltage:**
  - **Root pass:** 16–20 volts
  - **Fill and cap passes:** 18–22 volts (refer to the wire data sheet for precise settings)
✓ **Wire feed speed:**
  - **Root pass:** 10–16 ft/min (3–5 m/min)
  - **Remaining passes:** 13–30 ft/min (4–9 m/min)
✓ **Stick-out length:** 0.2–0.4″ (5–10mm) for short-circuit arc
✓ **Shielding gas flow rate:** None required (self-shielded)
✓ **Transfer mode:** Short-circuit
✓ **Bevel angle:** 35° V preparation
✓ **Additional considerations:** 1–2mm root face, 2.5mm root gap
IMPORTANT: May require reverse polarity (torch connected to negative).

ADVANTAGES OF SELF-SHIELDED FLUX-CORED WIRE FOR BUTT JOINTS COMPARED TO OTHER WIRES

1. **No need for shielding gas (compared to solid wire, metal-cored wire, rutile flux-cored wire, and basic flux-cored wire)**
   ⚠ Gas-free operation: Self-shielded flux-cored wire does not require shielding gas, making it ideal for outdoor welding, windy conditions, and remote locations where gas supply is impractical.
2. **Portability and simplicity of equipment (compared to solid wire, metal-cored wire, rutile flux-cored wire, and basic flux-cored wire)**
   ⚠ Lightweight and portable setup: No gas cylinders are needed, making the welding equipment more compact and easier to transport, which is particularly useful for on-site jobs or difficult-to-access areas.
3. **Higher tolerance for less rigorous surface preparation (compared to rutile flux-cored wire and solid wire)**
   ⚠ Better performance on contaminated surfaces: Self-shielded wire tolerates rust, oil, and other surface contaminants better, making it a good choice for fieldwork or industrial applications where surface cleaning may be limited.
4. **Greater adaptability to harsh environmental conditions (compared to basic flux-cored wire)**
   ⚠ Less sensitive to moisture and airflow: Self-shielded wire performs well even in high humidity and strong winds, maintaining arc stability in outdoor environments where gas-shielded wires struggle.

DISADVANTAGES OF SELF-SHIELDED FLUX-CORED WIRE FOR BUTT JOINTS COMPARED TO OTHER WIRES

1. **Higher fume and spatter generation (compared to solid wire and metal-cored wire)**
   ⚠ Increased smoke and spatter production: Self-shielded wire generates more fumes and spatter, requiring proper ventilation in confined spaces and making weld pool visibility more challenging.

2. **Lower weld bead appearance quality (compared to solid wire)**
   ⚠ Rougher weld finish: The weld bead tends to be less smooth and aesthetically pleasing, which may not be ideal for visible joints where appearance matters.
3. **Lower root penetration (compared to metal-cored wire and basic flux-cored wire)**
   ⚠ Weaker root fusion: Self-shielded wire has lower penetration capability, making it less suitable for thick butt joints or critical structural applications requiring full fusion.
4. **Denser slag and more cleaning effort (compared to rutile flux-cored wire)**
   ⚠ Slag is harder to remove: The slag produced by self-shielded wire is thicker and more difficult to remove, increasing cleaning time between passes and reducing overall productivity.
5. **Limited penetration and control in high-strength applications (compared to basic flux-cored wire)**
   ⚠ Not ideal for high-strength structural welding: While adequate for general applications, self-shielded wire lacks the deep penetration and mechanical strength of basic flux-cored wire, making it less suitable for high-stress joints.

COMMON DEFECTS IN MIG/MAG WELDING

MIG/MAG welding is a highly efficient process, but defects can sometimes appear in the weld beads, affecting the quality of the work. These issues can arise due to a variety of reasons, such as incorrect equipment settings, improper welding technique, or unfavourable environmental conditions. In the next section, we explain some of the most common defects encountered in MIG/MAG welding, along with tips on how to identify and correct them.

COMMON DEFECTS IN MIG/MAG WELDING

1. **Porosity**
   ✓ **What is it?**
   Porosity occurs when small gas bubbles become trapped inside the weld bead. These bubbles form when gases do not have enough time to escape before the molten metal solidifies.
   ✓ **Why does it happen?**
   - The workpiece surface or welding wire is contaminated.
   - Incorrect shielding gas flow rate.
   - Air currents displace the shielding gas.
   - The workpieces are damp or covered in rust.
   ✓ **How to prevent it?**
   - Thoroughly clean the base material and wire before welding.
   - Adjust the shielding gas flow rate according to the nozzle diameter and working conditions.
   - Avoid welding in windy areas; use screens to protect the welding zone if necessary.
   - Use wires designed to reduce oxidation when welding carbon steels.
2. **Undercut**
   ✓ **What is it?**
   Undercut appears as depressions or grooves along the edge of the weld bead where the base metal has been eroded or not properly filled.

✓ **Why does it happen?**
- The welding current is too high.
- The wire feed speed or torch travel speed is incorrect.
- The torch is positioned at an improper angle.

✓ **How to prevent it?**
- Reduce the welding current to avoid excessive melting of the base metal.
- Adjust the travel speed to ensure proper bead filling.
- Keep the torch at an appropriate angle (usually between 10° and 15°) to distribute the filler metal evenly.

3. **Slag inclusion**

✓ **What is it?**

Slag inclusions are non-metallic particles trapped in the weld bead. This issue is more common when using flux-cored wires that produce slag.

✓ **Why does it happen?**
- Insufficient cleaning between weld passes.
- Travelling too slowly, allowing slag to mix with the molten metal.
- Incorrect torch angle.

✓ **How to prevent it?**
- Clean each weld pass thoroughly before continuing.
- Increase travel speed to prevent slag from mixing with the weld pool.
- Maintain a torch angle that directs the slag away from the weld bead.

4. **Lack of fusion**

✓ **What is it?**

Lack of fusion occurs when the filler metal does not properly bond with the base metal or previous weld passes, resulting in a weak weld.

✓ **Why does it happen?**
- Insufficient welding current.
- Excessive travel speed.
- Improper torch angle.

✓ **How to prevent it?**
- Increase the welding current to ensure proper melting of the metals.
- Reduce the travel speed to allow enough time for fusion to occur.
- Adjust the torch angle according to the welding position.

PRACTICAL SOLUTIONS TO PREVENT AND CORRECT DEFECTS

1. **Adjusting welding parameters**
   ✓ **Current:** Ensure the current is properly set according to the material thickness and type of weld. Excessive current can cause undercut, while low current may result in lack of fusion.
   ✓ **Travel speed:** The speed of the torch and wire feed must be balanced with the current to ensure proper bead formation and avoid defects. Always keep ahead of the weld pool.
   ✓ **Gas flow rate:** Adjust the shielding gas flow according to the nozzle diameter and working conditions to ensure proper weld pool protection.

2. **Welding technique**
   ✓ **Torch angle:** Maintain the torch at the correct angle, typically 10° to 15°, to ensure even distribution of the filler metal.

✓ **Torch movement:** Use a steady, controlled motion to avoid fluctuations in the weld pool that could cause defects.

✓ **Cleaning between passes:** Remove slag and contaminants between weld passes to prevent inclusions and ensure good fusion between metal layers.

3. **Environmental control**

✓ **Wind screens:** If welding in an area with air currents, use screens to protect the welding zone and prevent the shielding gas from dispersing.

✓ **Preheating:** For thick materials or cold environments, preheating the workpiece can help prevent cracks and fusion issues.

4. **Equipment maintenance**

✓ **Contact tips and nozzle:** Keep contact tips and nozzles clean and in good condition to ensure a **consistent flow of current and gas**.

✓ **Rollers and wire feeder:** Ensure the rollers are properly adjusted to **prevent feeding issues** that could impact the welding process.

CONCLUSION

Identifying and correcting defects in MIG/MAG welding is essential for any welder. Understanding the causes of each defect and how to adjust parameters and technique to fix them not only improves weld quality but also saves time and reduces rework. With practice and attention to detail, you can minimise these problems and achieve strong, durable, high-quality welds.

"An expert is a person who has made all the mistakes possible in a given field."

—Niels Bohr

"Sometimes, it is the people no one imagines anything of who do the things that no one can imagine."

—Alan Turing

4 Oxygen Welding

PERSONAL GROWTH *"THE BAMBOO AND THE OAK"*

During a storm, a mighty oak tree was broken by the wind, while the bamboo remained standing, bending with each gust. The next day, the bamboo whispered to the fallen oak:

"Strength is not always about resisting, but about knowing how to adapt and grow with what life offers us."

Like the bamboo, grow in flexibility, and you will find strength in every step of your journey.

INTRODUCTION TO OXY-GAS WELDING

Oxy-gas welding consists of permanently joining two metal pieces by completely melting them with heat generated by a flame, which results from the combustion of a gas mixed with oxygen. It can be performed with or without a filler rod.

HISTORICAL ORIGINS OF OXY-GAS WELDING

Although oxy-gas welding is a well-established process, it remains widely used in:

- Thin sheet metal work.
- Welding pipes for low-pressure installations.
- Artistic metalwork.

This method requires skill, patience, and sensitivity to master, as it is not suited to hurried work.

Flame welding was the first welding process to be introduced into industry, once technology allowed the safe storage of the required gases. In 1901, the first oxy-acetylene torches were introduced, and by 1916, this process was already being used for welding steel, aluminium, brass, cast iron, and deoxidised copper.

FUNDAMENTALS OF OXY-GAS WELDING

Required equipment (see Figure 4.1):

- Fuel gas and oxygen cylinders.
- Special hoses for oxygen and fuel gas.
- Pressure regulators for each gas.
- Flashback arrestors for safety (see Figure 4.2).
- Cup-shaped spark lighter (flameless).
- Nozzle reamers for cleaning.
- Welding torch (or cutting torch).

DOI: 10.1201/9781003422488-5

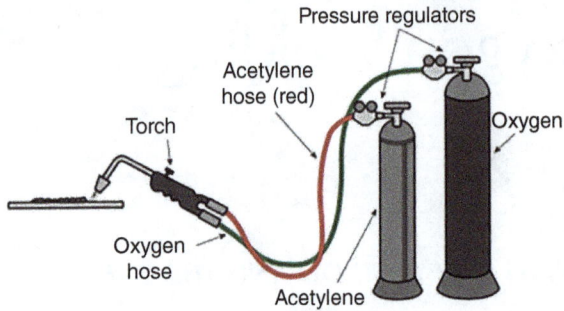

FIGURE 4.1 Diagram of an oxyacetylene welding setup with a torch, oxygen and acetylene hoses, gas cylinders, and pressure regulators.

FIGURE 4.2 Flashback arrestor safety device connected to an oxyacetylene welding gas hose.

- Nozzles with different flow rates.
- Filler rods.
- Welding shield or dark goggles specifically designed for gas welding.

GASES USED IN OXY-GAS WELDING

- **Oxygen:** This is the oxidizing gas that fuels combustion, increasing the flame temperature.
- **Fuel gas (mainly acetylene):**
  o Obtained from calcium carbide mixed with water, generating acetylene and lime as a residue.
  o One kilogram of carbide produces approximately 300 litres of acetylene.
  o Other gases like propane, butane, or propylene can also be used, though they create flames with different characteristics.

HOSES AND FLASHBACK ARRESTORS

- **Hoses:**
  o **Oxygen:** Blue or green, with right-hand thread.
  o **Acetylene:** Red, with left-hand thread.
  o Made of heat- and cut-resistant rubber.
- **Flashback arrestors:**
  o Prevent the flame from traveling back into the gas cylinders.
  o Installed at the regulator outlet and at the connection to the torch.

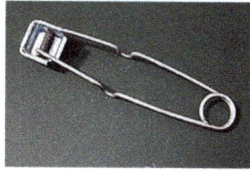

FIGURE 4.3   Cup-type spark lighter. Used to generate sparks that ignite the torch flame without the need for gas lighters.

FIGURE 4.4   Dismantled oxyacetylene torch, consisting of a handle with control valves and a nozzle with an interchangeable injector.

PRESSURE REGULATORS

- Each gas requires a specific regulator with different threads to prevent mix-ups.
- **Important:** Never grease a regulator or the cylinder thread! Contact between lubricant and oxygen or acetylene can be extremely dangerous.

CUP-TYPE SPARK LIGHTER

- Use spark lighters instead of gas lighters to avoid accidents (see Figure 4.3).

TORCH, NOZZLES, AND CLEANING TOOLS

- **Torch:**
  - o Two control valves: red for acetylene, blue or green for oxygen (see Figure 4.4).
- **Nozzles:**
  - o Interchangeable, available in different sizes (40–400L/H).
  - o Always keep them clean using the proper cleaning tools, and only when the nozzles are cold.

FILLER RODS

- **Diameters:** 1.6mm to 6mm.
- **Standard length:** 1 meter.
- May contain deoxidizers and alloying elements to improve welding properties.

WELDING HELMET/DARK GLASSES

- **Recommended shade level:** DIN 5–6.
- A full-face welding helmet is advisable for complete protection.

Step-by-step procedure:

1. **Inspection:**
   o   Check the hoses for cuts or burns.
   o   Open the gas cylinders and set the pressure according to the regulator instructions.

2. **Ignition:**
   o   Slightly open the acetylene valve and ignite using the spark lighter.
   o   Adjust the acetylene flow until a stable flame is obtained.

3. **Regulation:**
   o   Slowly open the oxygen valve until the flame forms a bright inner cone and a stable outer plume: **Neutral flame.**
   o   The **neutral flame** is ideal for welding steel, emitting CO_2 as a protective gas.

4. **Flame corrections:**
   o   **Carburizing flame:** Excess acetylene, used for welding cast iron.
   o   **Oxidizing flame:** Excess oxygen, used for welding brass.

5. **Cooling and cleaning:**
   o   If the nozzle overheats or clogs, cool it in a metal container with water.
   o   Use the appropriate cleaning tool to remove obstructions.

6. **Shutdown:**
   o   First, close the acetylene valve, then close the oxygen valve.

7. **Safety considerations:**
   o   While in use, ensure the torch is placed in a safe location, away from flammable materials.

Final note: The **neutral flame** reaches temperatures exceeding **3,000°C**, and its brightness can damage eyesight. Never underestimate its risks—**always wear proper protection.**

OXYACETYLENE WELDING

Practice 1. First beads in horizontal position (pa/1g)

- **Base material:** Two carbon steel plates, 4″ × 1.18″ × 0.059″ (100 × 30 × 1.5 mm).
- **Filler rod:** 1/16″ (1.6mm) for carbon steel.
- **Nozzle:** 0.06 cubic feet per hour (160 L/H).
- **Oxygen pressure:** As specified by the nozzle manufacturer.
- **Acetylene pressure:** As specified by the nozzle manufacturer.
- **Number of beads:** One, with a straight motion.
- **Welding direction:** Torch moving forward or backward.
- **Shade level of protective glasses or helmet:** No. 5.

1. **Material and equipment preparation**
   - Cut the plates to the indicated dimensions and file off any burrs from the cut (Figure 4.5).
   - Clean the plates with a cloth to remove grease or dirt residues.
   - Ensure that the torch nozzle is suitable for this exercise and check that the hoses have no cuts, cracks, or damage.
   - Open the gas cylinders and adjust the pressures on the regulators according to specifications.

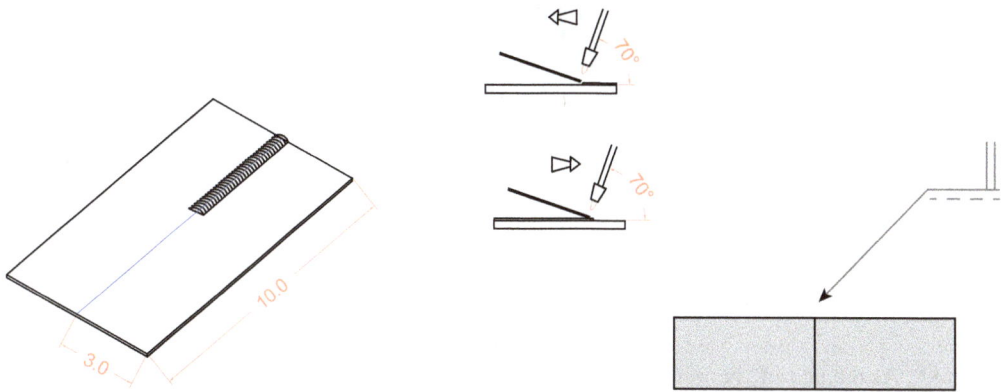

FIGURE 4.5 Illustration of a single-sided bevel butt weld on a metal sheet. The welding sequence and the recommended torch angle in different positions are shown. Reference lines indicate the dimensions of the piece and the welding travel direction.

2. **Ignition and flame adjustment**
   - Slightly open the acetylene valve and ignite the flame using the spark lighter.
   - Gradually open the oxygen valve until obtaining a **neutral flame**, characterized by a uniform bright cone.
   - Always keep the flame properly adjusted; if neutrality is lost (an excess of the plume appears), readjust the oxygen and acetylene.
3. **Tack welding the joint**
   - Clean the filler rods with a cloth before starting.
   - Align the plates edge to edge (no gap) and ensure they are perfectly straight.
   - Place five tack welds evenly along the joint to keep the plates in position.
4. **Welding technique**
   - Hold the torch at an angle between 20° and 40° to the plate, depending on the desired penetration.
   - Direct the inner cone of the flame at the joint, keeping it 0.2–0.4″ (5–10mm) from the surface (placing it too close may extinguish the flame).
   - When a small molten pool forms, gently introduce the filler rod into it. The rod tip should melt inside the pool, and when retracting it, ensure it remains within the protection of the flame.
5. **Welding motion**
   - Move slowly, ensuring the molten pool fully forms before advancing a few millimetres.
   - Add filler metal consistently to prevent excessive heat from burning through the plate. Wait to see sparks before adding filler metal.
   - If holes appear due to thermal expansion, add filler metal to fill them and ensure proper fusion.
6. **Adjustments during practice**
   - If the nozzle overheats and the flame extinguishes with a small explosion, close the acetylene valve, keep the oxygen open, and submerge the nozzle in a metal bucket with water to cool it down. Once cold, readjust the flame.
   - If the plates become too hot, reduce the acetylene and oxygen flow rates to work at a slower pace.

FIGURE 4.6   Steel plate with a weld bead in a vertical-up (PA) position, marked with chalk. The weld displays a uniform bead with well-defined ripples.

- At the end of the joint, tilt the torch backward gradually to reduce heat input and prevent the molten pool from breaking through the edge of the plate.
7. **Weld quality inspection**
   - Check the back side of the plate using pliers to avoid burns.
   - The weld should have completely penetrated, forming a small, even bead on the back side. If penetration is missing or the joint is not fully fused, review your technique.
8. **Overlapping welds in oxyacetylene welding**
   - When joining the end of one bead with the start of another, remelt the last 0.2″ (5mm) of the previous bead to achieve a seamless connection (Figure 4.6).

Important tips

1. **Keep the filler rod tip close to the bright cone**, inside the CO_2 protection generated by the neutral flame to avoid contamination.
2. If the flame loses its neutral state, **adjust the gas flow** rates immediately before continuing.
3. **Bend the last few centimetres of the filler rod** to prevent accidents; if you accidentally hit someone, it will reduce the impact force.
4. **Be patient.** At first, it may be challenging, but practice will help you develop confidence and control.

FINAL NOTE

Oxyacetylene welding requires **precision, care, and patience.** With **dedication and practice**, you will master this **"artistic"** welding process. **Don't give up—every attempt is a step closer to perfection!**

THE CHAINED ELEPHANT

Once upon a time, there was an elephant that, since childhood, had been tied to a stake with a chain. Though he tried to break free, the chain was too strong for him. Over time, he stopped trying, convinced that he would never be able to break it. Years later, as a powerful adult elephant, he remained tied to the same stake with the same chain. Even though he now had enough strength to break free, he never tried—because he believed it was impossible.

Often, the limitations we believe we have are merely a product of past experiences. Don't let limiting beliefs stop you from reaching your true potential.

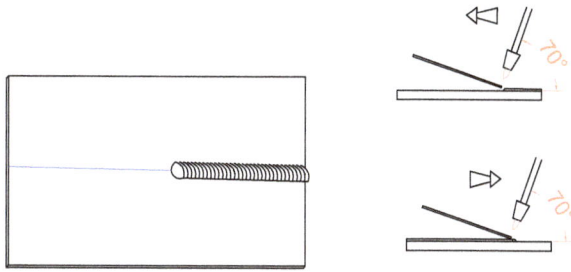

FIGURE 4.7   Butt joint weld using an oxyacetylene torch. The recommended torch angle is shown in two side views, with a 70° inclination in the travel direction and a lateral tilt adjusted for optimal molten pool control.

OXYACETYLENE WELDING

PRACTICE 2. FIRST BEADS IN HORIZONTAL VERTICAL POSITION (PC/2G)

- **Base material:** Two carbon steel plates, 4″ × 1.18″ × 0.059″ (100 × 30 × 1.5 mm).
- **Filler rod:** 1/16″ (1.6mm) for carbon steel.
- **Nozzle:** 0.06 cubic feet per hour (160 L/H).
- **Oxygen pressure:** As specified by the nozzle manufacturer.
- **Acetylene pressure:** As specified by the nozzle manufacturer.
- **Number of beads:** One, with a straight motion.
- **Welding direction:** Torch moving forward or backward.
- **Shade level of protective glasses or helmet:** No. 5.

GENERAL CONSIDERATIONS

The horizontal-vertical (PC/2G) position presents additional challenges when welding with oxy-acetylene due to the heat distribution and torch orientation. The heat generated by the flame tends to concentrate on the upper plate, which can be risky if not properly controlled (Figure 4.7).

- **Flame safety:** Always keep the flame directed toward the joint and away from your hands.
- **Hand movement:** Practice moving your hand up along the filler rod after depositing material three or four times to keep it away from the heat zone.

1. **Welding technique**
   1.1 Torch angle
      - Keep the torch tilted between **20° and 40°**, directing the heat toward the joint between the plates.
      - Avoid tilting the torch upwards, as this will overheat the upper plate and make controlling the molten pool more difficult.
   1.2 Torch movement
      - Weld forward, from right to left if you are right-handed.
      - Keep the flame cone **0.2–0.4″ (0.5–1cm)** from the joint, as this is the hottest part of the flame and allows optimal control of the molten pool.
   1.3 Filler rod application
      - Insert the tip of the filler rod into the molten pool, allowing it to fuse and deposit material through the joint.

- Add material lightly and consistently, ensuring that the molten metal flows toward the back to create a **small, uniform penetration bead**.

2. **Safety and control**
   - **Heat control:**
     - o The heat from the oxyacetylene flame tends to concentrate more on the upper plate due to the natural direction of rising heat. Avoid bringing your hand too close to the upper joint area.
   - **Hand positioning:**
     - o Move your hand up along the filler rod after depositing three or four times, keeping a safe distance from the heat zone.
   - **Flame cone distance:**
     - o Ensure you keep the flame cone at a consistent **0.2–0.4″ (0.5–1cm)** from the joint. If the flame is too close, it may extinguish or overheat the material, making molten pool control more difficult.

3. **Additional tips**
   - Practice maintaining a constant speed as you advance the torch to avoid irregularities in the bead.
   - If you notice that the flame loses neutrality, adjust the oxygen and acetylene flow rates until it stabilizes before continuing.
   - If the torch tip overheats, follow the proper safety cooling procedure:
     1. Turn off the acetylene valve.
     2. Keep the oxygen valve open and submerge the tip and torch neck in a metal bucket filled with water.
     3. Wait for it to cool completely before reigniting and adjusting the flame.

FINAL NOTE

Welding in the **PC/2G position** requires more precise torch control and good coordination between material deposition and torch movement. With practice, you will master this technique and achieve **a clean bead with uniform penetration**.

OXYACETYLENE WELDING

PRACTICE 3. FIRST BEADS IN VERTICAL-UP POSITION (PF/3G)

- **Base material:** Two mild steel plates, dimensions 4″ × 1.2″ × 1/16″ (100 mm × 30 mm × 1.5 mm)
- **Filler rod:** Mild steel rod, diameter 1/16″ (1.6mm)
- **Nozzle size:** 160 L/H (consult nozzle manufacturer's instructions)
- **Oxygen pressure:** Set according to nozzle manufacturer's recommendations
- **Acetylene pressure:** Set according to nozzle manufacturer's recommendations.
- **Number of beads:** Single bead with straight torch movement
- **Direction of torch travel:** Torch pointing forward or slightly backward (dragging)
- **Lens shade number (eye protection):** Shade N° 5

SPECIFIC TECHNIQUE FOR VERTICAL-UP WELDING

1. **Flame adjustment:**
   Set a neutral flame, slightly reducing the gas flow to weld more slowly and calmly, giving better control of the weld pool.

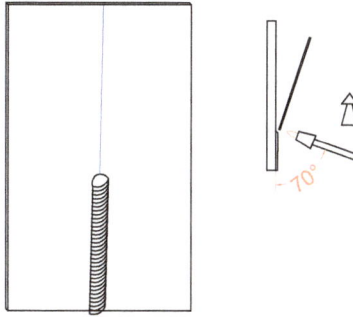

FIGURE 4.8   Upward vertical oxyacetylene weld. Shows a butt joint with a deposited bead and the recommended 10° torch angle.

2. **Starting the joint:**
   - Tilt the torch between **20° and 40°** and direct heat onto the starting point of the joint.
   - Once the metal surface changes colour (indicating sufficient heat), add a small amount of filler rod to protect the edges and avoid excessive melting.
3. **Progressing upward:**
   - Move upward about **1/8″ (3–4mm)** after the initial filler addition. Wait until the weld pool is sufficiently molten before adding more filler rod.
   - If the pool isn't molten enough, the rod may stick to the workpiece.
4. **Filler rod feeding and penetration:**
   - Dip the filler rod gently into the molten pool, applying slight pressure to ensure penetration through the joint to the opposite side.
   - Maintain the inner cone (dart) of the flame about **1/4–3/8″ (6–10mm)** away from the joint and carefully control melting to avoid defects.
5. **Heat precautions:**
   - Because heat rises, regularly reposition your hand on the filler rod, keeping it safely away from the hot weld pool.
   - Rest your elbow against your body for better stability and torch control.

Final advice:

Vertical-up welding requires patience and precision. Practise slowly, adjust flame intensity and positioning as necessary, and aim for a smooth, even bead with good penetration (Figure 4.8).

OXYACETYLENE WELDING

PRACTICE 4. FILLET WELD IN HORIZONTAL POSITION (PB/2F)

- **Base material:** Two mild steel plates, dimensions 4″ × 1.2″ × 1/8″ (100 mm × 30 mm × 3 mm)
- **Filler rod:** Mild steel rod, diameter 5/64″ or 3/32″ (2mm or 2.4mm)
- **Nozzle size:** 300 L/H (consult nozzle manufacturer's instructions)
- **Oxygen pressure:** Set according to nozzle manufacturer's recommendations
- **Acetylene pressure:** Set according to nozzle manufacturer's recommendations
- **Number of beads:** Single bead with straight torch movement

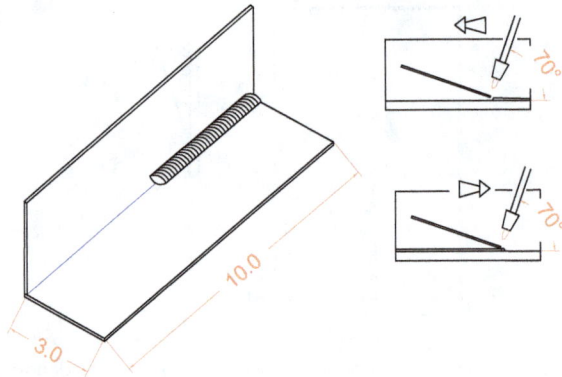

FIGURE 4.9 Oxyacetylene fillet weld. Shows a T-joint with a deposited bead and the recommended 10° torch angle.

- **Direction of torch travel:** Torch pointing forward or slightly backward (dragging)
- **Lens shade number (eye protection):** Shade N° 6

SPECIFIC CONSIDERATIONS FOR PB POSITION (HORIZONTAL FILLET WELD)

1. **Tack welding:**
   - Place a tack weld at each end of the joint (on the side to be welded). This will keep the plates stable.
   - Ensure the plates are properly aligned and have no gap between them.
2. **Controlling initial heat:**
   - In this position, it is particularly important to prevent excessive heat from melting the plate edges.
   - Keep the inner cone (dart) of the flame **1/4–3/8″ (6–10mm)** away from the joint.
   - Add a small amount of filler rod as soon as the metal begins to heat up, ensuring the fusion is concentrated on that drop.
3. **Torch angle:**
   - The torch should be tilted as follows:
     o **20° to 45°** in the direction of travel.
     o **45°** relative to both plates, ensuring even heat distribution.
4. **Filler addition and weld pool control:**
   - Wait until the weld pool is fully formed before adding filler metal. Insert the rod at the **highest point of the pool** to fill any undercut.
   - Move forward only when the pool is stable and well-developed.
5. **Correcting defects:**
   - If holes appear, try to fill them calmly. It is better to leave minor imperfections in this practice than to create excessive build-up due to hesitation in working the weld pool properly.
   - Learn to recognise the moment when the plate reaches its heat limit to improve future welds.
6. **Achieving a clean, narrow bead:**
   - To obtain a narrow, uniform bead, use only the minimum amount of filler rod required. Add material only when the weld pool starts to recede at the leading edge.

Final advice:

The key to this practice is finding the right balance between heat distribution and filler addition to achieve a clean, narrow weld bead. Work patiently and precisely, avoiding excessive material build-up that could affect the final weld quality (Figure 4.9).

OXYACETYLENE WELDING

PRACTICE 5. FILLET WELD IN VERTICAL-UP POSITION (PF/3F)

Base material: Two mild steel plates, dimensions 4″ × 1.2″ × 1/8″ (100 mm × 30 mm × 3 mm)
Filler rod: Mild steel rod, diameter 5/64″ or 3/32″ (2mm or 2.4mm)
Nozzle size: 300 L/H (consult nozzle manufacturer's instructions)
Oxygen pressure: Set according to nozzle manufacturer's recommendations
Acetylene pressure: Set according to nozzle manufacturer's recommendations
Number of beads: Single bead with straight torch movement
Direction of torch travel: Torch pointing forward
Lens shade number (eye protection): Shade N° 6

SPECIFIC CONSIDERATIONS FOR PF POSITION (VERTICAL-UP FILLET WELD)

1. **Positioning and visibility:**
   - Find a stance that allows you a clear view of the weld pool, preventing slow or inaccurate movements that could cause holes.
   - If you are right-handed, position your head in the upper left corner of the workpiece for an optimal view of the joint. However, feel free to experiment to find a more comfortable position.
2. **Controlling the neutral flame:**
   - In this position, involuntary torch movements can cause the neutral flame to become unstable.
   - Adjust the flame as often as necessary to maintain good fusion and avoid defects.
3. **Cleaning and maintenance:**
   - Cleaning both the plates and the nozzle remains crucial in this practice to ensure a consistent weld pool and a clean finish.

Final advice

Success in this practice depends on steady torch movement and the ability to maintain constant visibility of the weld pool. Do not rush—adjust your position as needed to achieve a uniform bead (Figure 4.10).

OXYACETYLENE WELDING

PRACTICE 6. 2-INCH PIPE ON HORIZONTAL PLATE (PB/2F POSITION)

- **Base material:**
  Steel plate: 4″ × 4″ × 1/8″ (100 mm × 100 mm × 3 mm)
  Steel pipe: 2″ (50.8mm) diameter × 1.2″ (30mm) length × 1/8″ (3mm) wall thickness

- **Filler rod:** Mild steel rod, diameter 5/64″ or 3/32″ (2mm or 2.4mm)
- **Nozzle size:** 300 L/H (consult nozzle manufacturer's instructions)
- **Oxygen pressure:** Set according to nozzle manufacturer's recommendations
- **Acetylene pressure:** Set according to nozzle manufacturer's recommendations
- **Number of beads:** Single bead with straight torch movement
- **Direction of torch travel:** Torch pointing forward or backward (dragging or pushing)
- **Lens shade number (eye protection):** Shade N° 6

STEP-BY-STEP WELDING PROCEDURE

1. **Joint preparation:**
   - Carefully clean the welding area on both the pipe and the plate to remove rust, grease, or impurities.
   - Position the pipe on the plate in a horizontal position and tack-weld it at four points, one at each corner of the joint, as shown in the diagram.
2. **Welding technique:**
   - Adjust the nozzle and the oxygen and acetylene flow rates according to the recommended parameters.
   - Keep the inner cone (dart) of the flame 1/4–3/8″ (6–10mm) away from the joint.
   - Start the weld at one end, ensuring the molten pool is well-controlled, allowing it to penetrate slightly into both the pipe and the plate.
   - Add filler rod moderately when the molten pool is well-formed. The filler should slightly increase the level of the pool before moving forward.
3. **Welding in sections:**
   - Divide the weld bead into four or five sections, alternating sides to minimise distortions.
   - When restarting each section, remelt the last 1/4″ (5mm) of the previous weld bead to ensure proper continuity.
4. **Additional tip:**
   - If you find it difficult to coordinate the torch hand with the filler rod, perform an initial pass without filler to get familiar with moving along the curved surface of the pipe.

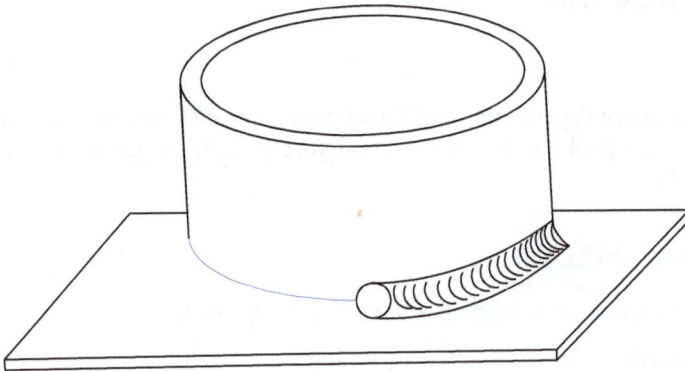

FIGURE 4.10 Oxyacetylene butt weld of a pipe to a plate. Shows the start of the weld bead around the pipe.

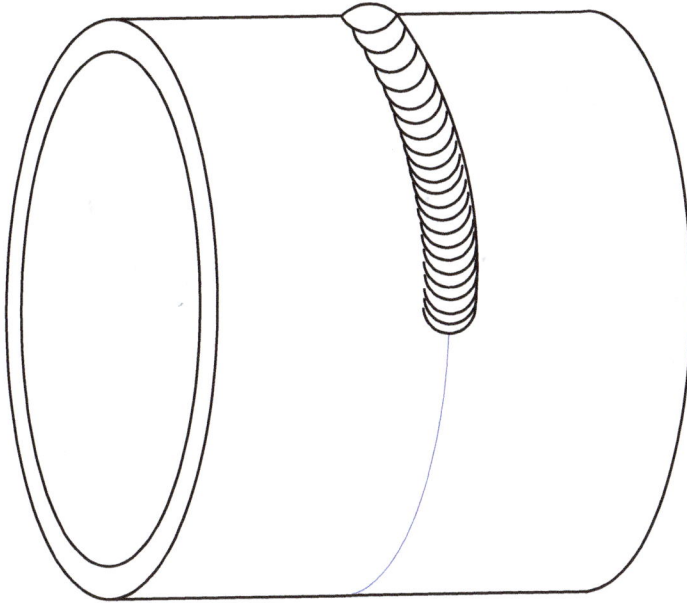

FIGURE 4.11 Oxyacetylene butt weld on a pipe. Shows the start of the weld bead at the pipe joint.

Practical tips:

✓ **Ensure good visibility** when starting each section to maintain proper control of the weld pool.

✓ **Adopt a comfortable and stable position** to facilitate work on the curved surface.

✓ **Avoid overheating the joint,** as excessive heat can cause excessive deformation or welding defects (Figure 4.11).

OXYACETYLENE WELDING

PRACTICE 7. HORIZONTAL PIPE JOINT WITH ROTATION (PA/1G POSITION)

- **Base material:** Two carbon steel pipes, dimensions 2″ (50.8mm) diameter × 1.2″ (30mm) length × 1/8″ (3mm) wall thickness
- **Filler rod:** Mild steel rod, diameter 5/64″ or 3/32″ (2mm or 2.4mm)
- **Welding equipment:** Nozzle size 300 L/H
- **Oxygen pressure:** Set according to nozzle manufacturer's recommendations
- **Acetylene pressure:** Set according to nozzle manufacturer's recommendations
- **Number of beads:** Single bead with straight torch movement
- **Direction of torch travel:** Torch pointing forward
- **Lens shade number (eye protection):** Shade N° 6

STEP-BY-STEP WELDING PROCEDURE

1. **Pipe preparation:**
   - Cut, remove burrs, and thoroughly clean the pipe ends.
   - Align the pipes without any gap and tack-weld them at four points (one in each quadrant).

2. **Positioning:**
   - Lay the pipes horizontally on a stable surface.
   - Use a support to prevent rolling during welding.
3. **Welding technique:**
   - Start welding at the 3 o'clock position and move towards 12 o'clock in an anticlockwise direction.
   - Tilt the torch between 20° and 45°, keeping the inner cone (dart) 1/4–3/8″ (6–10mm) from the joint.
   - Add filler only to the sides of the joint, alternating continuously and ensuring the molten pool is sufficiently hot before each addition (you'll know it's ready when small sparks appear from the weld pool).
4. **Welding sequence:**
   - Divide the bead into four sections, alternating sides to minimise distortion.
   - Weld the following sections from 3 o'clock to 12 o'clock.
   - Remelt the last 1/4″ (5mm) of each previous bead to ensure uniform fusion at the overlap points.
5. **Final test:**
   - Allow the weld to cool completely before inspection.
   - In a gas installer certification exam, this exercise is tested with a pressure test to detect porosity.

PRACTICAL TIP

Maintain a smooth and consistent torch movement, adjusting speed according to your skill level to ensure a high-quality bead. The key is to coordinate edge fusion and filler addition on the sides to achieve proper penetration.

OXYACETYLENE WELDING

PRACTICE 8. PIPE JOINT IN OVERHEAD POSITION (PC/2G)

- **Base material:** Three carbon steel pipes, dimensions 2″ (50.8mm) diameter × 1.2″ (30mm) length × 1/16″ (1.5mm) wall thickness
- **Filler rod:** Mild steel rod, diameter 5/64″ or 3/32″ (2mm or 2.4mm)
- **Welding equipment:** Nozzle size 300 L/H
- **Oxygen pressure:** Set according to nozzle manufacturer's recommendations
- **Acetylene pressure:** Set according to nozzle manufacturer's recommendations
- **Number of beads:** Single bead with straight torch movement
- **Direction of torch travel:** Torch pointing forward
- **Lens shade number (eye protection):** Shade N° 6

STEP-BY-STEP WELDING PROCEDURE

1. **Pipe preparation:**
   - Clean the pipe ends and tack-weld all three pipes together, stacking them one on top of the other.
   - Depending on the results of previous exercises, you may or may not leave a 2mm gap between the pipes.
2. **Flame adjustment:**

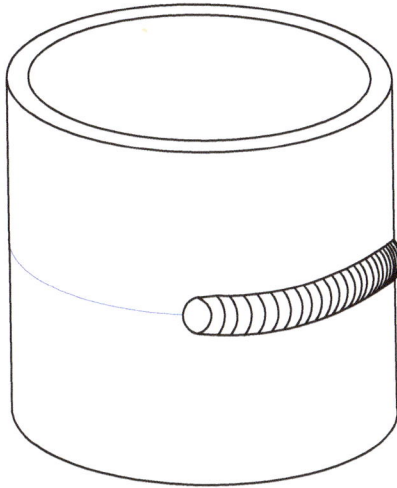

FIGURE 4.12   Circumferential weld on a pipe using oxyacetylene. Shows the progression of the weld bead around the joint.

- Set a neutral flame to ensure a clean and controlled fusion.
3. **Welding technique:**
   - Weld each joint in four sections (quarters), alternating between sides and pipes to distribute heat evenly.
   - Start with the torch tilted at 45° relative to the joint, keeping the inner cone (dart) 1/4–3/8″ (6–10mm) from the molten pool.
   - Add filler continuously when the molten pool is well-formed, ensuring complete fusion of the edges.
4. **Alternating sides:**
   - Once you complete a quarter of a joint, switch to the opposite side of the other joint.
   - This helps balance expansion and contraction across the structure.
5. **Overlapping Beads:**
   - At overlap points, remelt the last 1/4″ (5mm) of the previous weld bead to ensure seamless continuity (Figure 4.12).

PRACTICAL TIP

Even heat distribution across the pipes is crucial to avoid deformations. Work at a steady pace and monitor contraction as the weld cools. If the final assembly remains straight, you have achieved a precise and professional weld.

COMMON DEFECTS IN OXYACETYLENE WELDING

1. Porosity

What is it?

Porosity appears as small gas bubbles trapped within or on the surface of the weld bead. These bubbles weaken the joint and affect the weld's appearance.

Why does it happen?

- Contaminated base material or filler rod (rust, grease, or moisture).
- Excess gases in the oxygen-fuel mixture.
- Poor welding technique allowing air to enter the molten pool.

How to prevent it?

- ✓ Clean the base material and filler rod carefully before welding.
- ✓ Adjust the gas mixture correctly to avoid excess oxygen or fuel.
- ✓ Maintain a steady technique that prevents air from entering the molten pool.

2. Lack of fusion

What is it?

Lack of fusion occurs when the filler metal fails to bond properly with the base material or previous passes, resulting in a weak joint.

Why does it happen?

- Flame temperature too low due to incorrect gas adjustment.
- Advancing too fast with the filler rod or torch.
- Incorrect torch positioning in relation to the base material.

How to prevent it?

- ✓ Adjust the gas mixture to obtain a neutral flame with sufficient heat.
- ✓ Slow down to allow the flame enough time to melt and fuse the materials properly.
- ✓ Keep the flame focused on the work area at the correct angle to improve fusion.

3. Oxide inclusions

What is it?

Oxide inclusions are particles trapped within the weld bead, originating from oxidised surfaces or improper flame combustion.

Why does it happen?

- Contaminated or oxidised base material or filler rod.
- Use of an oxidising flame (excess oxygen).
- Prolonged exposure of the molten pool to air.

How to prevent it?

- ✓ Thoroughly clean the base material and filler rod before welding.
- ✓ Adjust the gas mixture to maintain a neutral flame, avoiding excess oxygen.
- ✓ Work quickly to minimise molten pool exposure to air.

4. Overfusion

What is it?

Overfusion occurs when the base material melts excessively, creating weak zones or even perforations in the metal.

Why does it happen?

- Excessively hot flame due to too much oxygen or fuel.
- Keeping the flame on the same spot for too long.
- Slow torch movement, concentrating heat in one area.

How to prevent it?

- ✓ Adjust the gas mixture to maintain a neutral flame with controlled heat.
- ✓ Keep the torch in constant motion to prevent excessive heat build-up.
- ✓ Slightly increase torch travel speed when working with thin materials.

5. Irregular weld bead

What is it?

An irregular weld bead shows variations in width, height, or shape, affecting both appearance and strength.

Why does it happen?

- Inconsistent torch or filler rod movements.
- Improper torch travel speed.
- Fluctuations in the gas mixture during welding.

How to prevent it?

- ✓ Practise steady and uniform movements with both the torch and filler rod.
- ✓ Adjust travel speed according to material thickness.
- ✓ Check welding equipment to ensure a steady gas supply.

PRACTICAL SOLUTIONS TO PREVENT AND CORRECT DEFECTS

1. **Welding parameter adjustments**
   - **Gas mixture:** Configure the correct ratio of oxygen and fuel gas (typically acetylene or propane) to maintain a neutral flame that avoids oxidation or excess heat.
   - **Temperature:** Adjust flame intensity according to base material thickness.
2. **Welding technique**
   - **Torch positioning:** Keep the flame at an optimal angle (typically 45–60°) to direct heat efficiently into the weld joint.

- **Consistent motion:** Use steady, controlled movements to maintain a uniform molten pool and avoid irregular beads.
- **Surface preparation:** Remove all contaminants from both the base material and filler rod before welding.

3. **Environmental control**
   - Work in a clean area, protected from drafts that could affect the flame or introduce contaminants into the molten pool.
   - Preheat materials if working in cold conditions or with thicker sections, improving molten pool fluidity.

4. **Equipment maintenance**
   - Regularly inspect torch nozzles to ensure they are clean and in good condition.
   - Check hoses and connections to prevent gas leaks or pressure fluctuations.

CONCLUSION

Oxyacetylene welding is a versatile and effective process that requires precision and careful technique to avoid common defects. Adjusting parameters, maintaining proper technique, and working in a controlled environment are key to achieving high-quality welds. With practice and attention to detail, it is possible to master this process and produce consistent, reliable welds.

THE HUMMINGBIRD'S PARABLE

A forest was on fire, and while all the animals fled to save their own skin, a little hummingbird kept flying back and forth, carrying drops of water from the river to pour onto the flames.

"Do you really think that with such a tiny beak you're going to put out the fire?" asked the lion.

"I know I can't do it alone," the little bird replied, "but I'm doing my part."

"I understand, but when I act like you, I feel powerless, as if it makes no difference. Sometimes I wish others would behave differently, that things were not the way they are. I can't help but criticise certain actions or people, and it makes me suffer."

"Then live like the flowers!"

"And what does it mean to live like the flowers?"

"Pay attention. Do you see those flowers growing in the garden? They bloom in manure, yet they remain pure and fragrant. They take from the foul-smelling soil only what is useful and nourishing, but they do not allow the bitterness of the earth to stain the freshness of their petals. It is fair to be troubled by one's own faults, but it is unwise to let the flaws of others disturb you. Their defects belong to them, not to you. And if they are not yours, why let them trouble you? Practise the virtue of rejecting all the harm that comes from outside, and perfume the lives of others by doing good. That is what it means to live like the flowers."

Yoga. El silencio es mi alimento —Vicente Moreno

Bibliography

BOOKS

- **American Welding Society** (2015) *Welding Handbook: Welding Processes, Part 1 (Vol. 2)*. Miami, FL: AWS.
- **Blunt, J. and Balchin, N.** (2002) *Welding: Principles and Applications* (6th ed.). Albany, NY: Delmar Cengage Learning.
- **Cary, H.B. and Helzer, S.C.** (2005) *Modern Welding Technology* (6th ed.). Upper Saddle River, NJ: Pearson Education.
- **Hicks, J.** (2014) *Welded Joint Design*. Cambridge: Woodhead Publishing.
- **International Institute of Welding (IIW)** (2018) *Guide for the Application of ISO 3834: Quality Control in Fusion Welding*. Geneva: IIW.
- **Kalpakjian, S. and Schmid, S.** (2020) *Manufacturing Processes for Engineering Materials* (7th ed.). Pearson Education.
- **Rampaul, H.** (2003) *Pipe Welding Procedures* (3rd ed.). New York: Industrial Press.

SCIENTIFIC ARTICLES

- **Almeida, A. and Quintino, L.** (2010) 'Welding Defects and Their Control', *International Journal of Advanced Manufacturing Technology*, 50(1), pp. 29–41.
- **Shi, Y. and Song, G.** (2012) 'Effect of Welding Parameters on Porosity Formation in TIG Welding', *Journal of Materials Processing Technology*, 212(9), pp. 2041–2047.

TECHNICAL STANDARDS & PUBLICATIONS

- **American Welding Society (AWS)** (2017) *AWS D1.1/D1.1M: Structural Welding Code – Steel*. Miami, FL: AWS.
- **International Organization for Standardization (ISO)** (2018) *ISO 3834–2: Quality Requirements for Fusion Welding of Metallic Materials*. Geneva: ISO.
- **European Committee for Standardization (CEN)** (2019) *EN 1011–1: Recommendations for Welding of Metallic Materials*. Brussels: CEN.

WEBSITES

- **Miller Electric** (n.d.) 'Resources and Articles on Welding Processes'. Available at: www.miller-welds.com/resources (Accessed: [date]).
- **Lincoln Electric** (n.d.) 'Welding Knowledge Center'. Available at: www.lincolnelectric.com/en-us/support/welding-how-to (Accessed: [date]).
- **Welding tips and tricks** (n.d.) 'Educational Videos and Tutorials'. Available at: www.weldingtipsandtricks.com/ (Accessed: [date]).

VIDEO RESOURCES & MULTIMEDIA

- **Facultad de Soldadura** (n.d.) 'YouTube Channel'. Available at: www.youtube.com/@Facultaddesoldadura (Accessed: [date]).

Index

For Product Safety Concerns and Information please contact our EU
representative GPSR@taylorandfrancis.com
Taylor & Francis Verlag GmbH, Kaufingerstraße 24, 80331 München, Germany

www.ingramcontent.com/pod-product-compliance
Lightning Source LLC
Chambersburg PA
CBHW061347210326
41598CB00035B/5905

9 781032 692265